The Seven Circles Theorem

and other new theorems

by

C. J. A. Evelyn

G. B. Money-Coutts

J. A. Tyrrell

First published 1974

ISBN 0950 3304 0X

This book has been printed by letterpress and bound in Great Britain for the publishers, Stacey International (Registered Office: 128 Kensington Church Street, London W8).

Distributed by Business Books Limited, Mercury House, Waterloo Road, London SE1.

MADE AND PRINTED IN GREAT BRITAIN

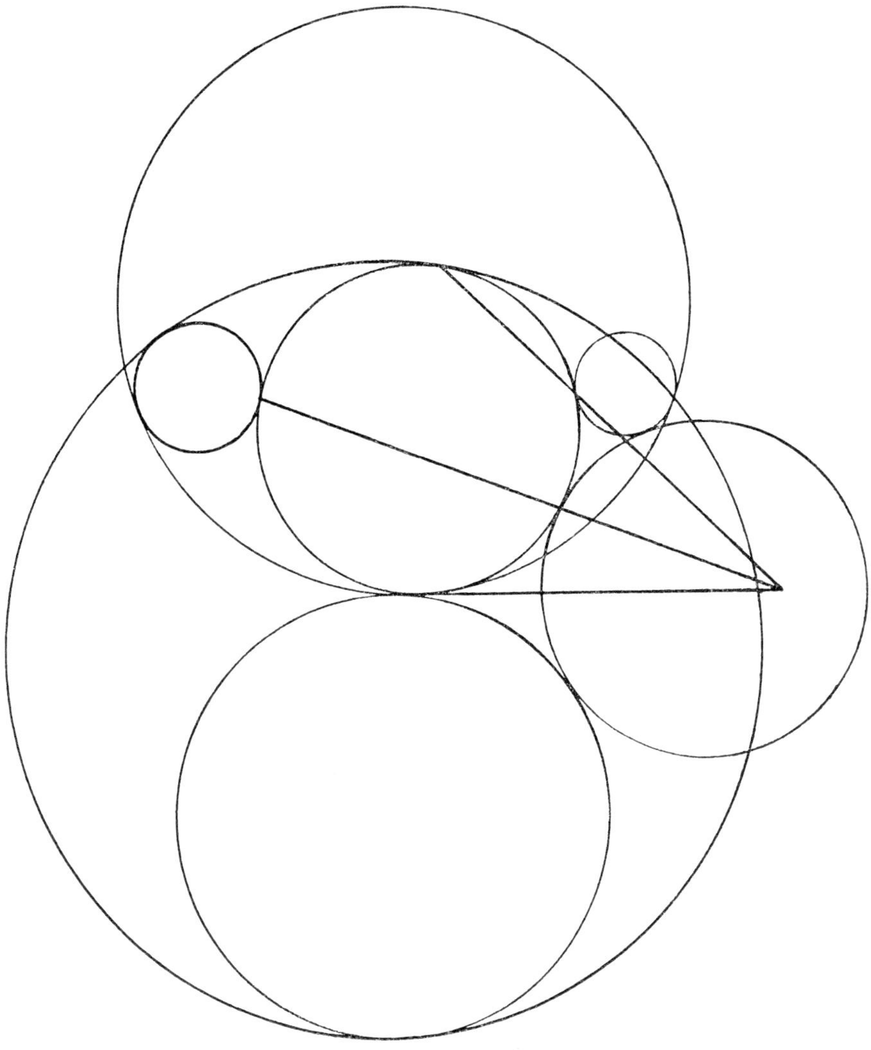

THE

SEVEN CIRCLES THEOREM

(See Section 3.1)

Printed by
The Elvy and Gibbs Partnership,
11 Best Lane, Canterbury, Kent.

CONTENTS

PREFACE

This collection of theorems in plane geometry, some of which we believe to be new, illustrates that there may still be plenty of interesting theorems awaiting discovery. We derived a lot of enjoyment from evolving them (or, as it transpired in some cases, rediscovering them) and it is our sincere hope that the reader may derive as much pleasure from reading about them. We have included a rather more generous proportion of diagrams than is strictly necessary, principally, of course, to assist the reader in following the proofs, but partly, too, for their beauty of design, which we have found to have an appeal to many of our friends who are not mathematicians.

The book is arranged in three main sections, the first on a topic in linear geometry, the second on some generalisations of Pascal's theorem, and the third on a series of theorems concerning chains of circles (none of which, surprisingly, appear to be known). Most of the proofs given are essentially elementary and should be within the comprehension of an average university student in mathematics.

We wish to express our gratitude to the late Dr. T. W. Chaundy for the interest he showed in the theorems and particularly for his elegant proof of the Seven Circles Theorem (given in Section 3.1). We also thank Mr. Robert Cordner, who supplied the proof of the Heptagon Theorem (Section 2.1); Dr. Patrick Du Val, for clarifying a number of points in connection with Section 2; and Mr. J. F. B. Wardale, who contributed the proofs of the theorems in Sections 3.2 and 3.3.

33 Grosvenor Square,
London, W.1.
November, 1971

C.J.A.E.
G.B.M.-C.
J.A.T.

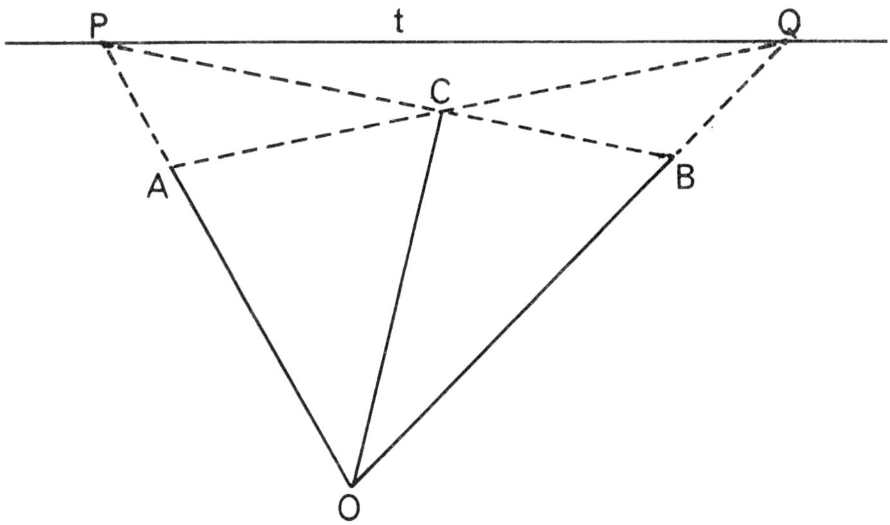

Figure 1a (see Section 1.1)

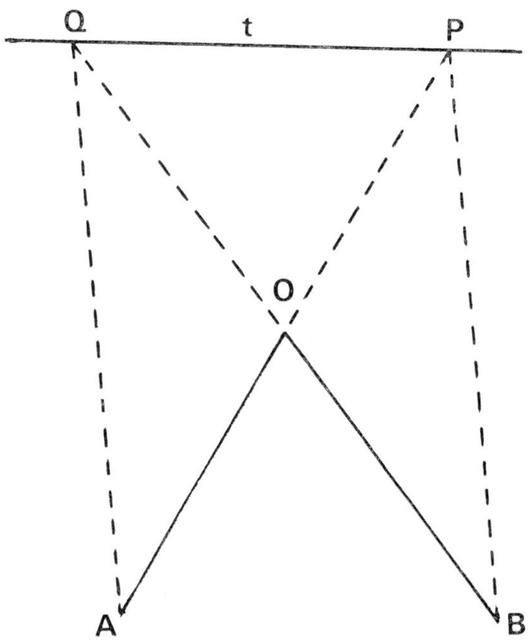

Figure 1b (see Section 1.1)

SECTION 1

PROJECTIVE ADDITION OF VECTORS AND THE DUAL THEORY

Some fifty years ago, L. Silberstein wrote an interesting little book entitled 'Projective Vector Algebra' (Bell, London, 1919) of which the aim was to develop an algebra of vectors independent of the axioms of congruence and parallelism. A great deal of this book is concerned with the construction and elucidation of what we should now call an unusual vector space structure on a slightly modified Euclidean plane. Unlike most vector space structures, however, the verification of the axioms is not just a trivial matter; some fairly advanced geometry is required. Moreover—and this is the main charm of Silberstein's book—the development of the construction lends itself well to a 'discovery' method of approach and—or so we think—the geometrical theory and the vector space theory are made to complement one another in a way that may, even today, have didactic appeal to teachers of geometry and algebra at the advanced sixth form or first year University level. We have therefore thought it worth while to devote this first section to a brief summary of Silberstein's discussion in the hope that it may prove of some use to teachers who do not have access to his book. We have also taken the opportunity to add some indications of the dual theory.

1.1. POINT-POINT VECTORS

A *vector* is just an ordered pair of points (A, B) in the plane (or in space), A being called its *initial point* and B its *end-point*; in accordance with common practice, we use the symbol AB to denote this vector. Vectors of this kind (which may be described more fully as point-point vectors) are no different from those with which we ordinarily deal; the novelty lies in the manner of adding two of them together, which we now describe.

We select, quite arbitrarily, a line t in the plane (or a plane t if we are concerned with three-dimensional space) and we define addition of vectors *relative to t*. Consider first two vectors OA and OB which share a common initial point O (not on t) and which do not lie in the same direction (i.e. O, A and B are not collinear). We let P and Q be the points of intersection of t with the lines OA and OB respectively (produced if necessary) and we let C be the point of intersection of the lines AQ and BP (cf. Figure 1a). We then define *the projective sum of OA and OB relative to t* to be the vector OC and we write

$$OA + OB = OC$$

or, if we wish to emphasize the role of t,

$$OA + OB = OC \text{ (rel } t\text{)}.$$

Two points of interest arise out of this definition:

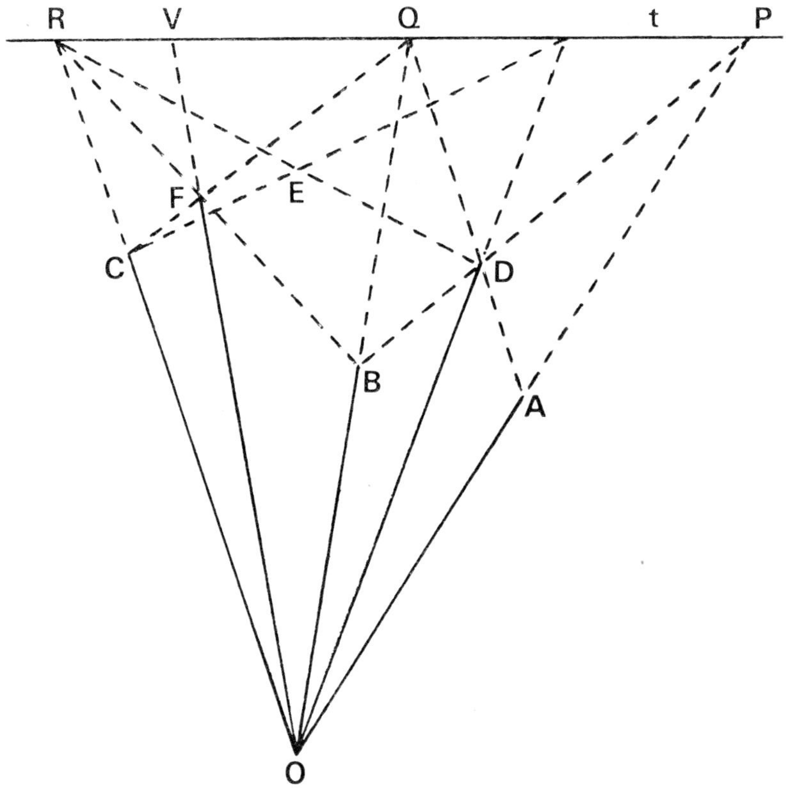

Figure 2 (see Section 1.1)

(i) If both A and B lie on t, the above construction for C becomes indeterminate and (if only by considering nearby positions of A and B) it soon becomes clear that no sensible meaning can be ascribed to $OA + OB$. Arguing from this case, one easily discovers that, in order to have a well-defined operation of projective addition (relative to t), *it is necessary to exclude from consideration all vectors with end-points on t.*

(ii) It can also happen (cf. Figure 1b) that the lines AQ and BP are parallel, so that C is at infinity; and discussion of this case leads to the necessity of including, as part of our system, *ideal vectors* OC with C at infinity (in the sense of projective geometry).

If t is taken to be the line (or plane) at infinity, the exceptional cases (i) and (ii) do not occur and addition relative to t reduces to the ordinary 'parallelogram' method of adding vectors. Of course, the advanced student will realise that the properties of addition relative to t can easily be obtained by projecting t to infinity and making appeal to the properties of ordinary vector addition. But, for the less advanced student, there are many useful lessons to be learned by pursuing the matter at a less sophisticated level. Thus, leaving aside the commutative law of addition (which is obvious from the definition) the question arises as to whether addition relative to t is associative, i.e. whether

$$(OA + OB) + OC = OA + (OB + OC) \qquad (1)$$

whenever both sides are defined. This is by no means obvious and requires for its proof the theorem of Desargues on perspective triangles (or an equivalent result). Thus, in Figure 2, OD represents $OA + OB$, OE represents $OD + OC$ and OF represents $OB + OC$; and, to prove the associative law (1), it is necessary to show that $OE = OA + OF$, i.e. that the lines AV and FP intersect at E. This is achieved by applying the theorem of Desargues first to triangles CFE and OBD and then to triangles ADE and OBF.

Once associativity is established, one can turn one's attention to the detail of defining $OA + OB$ when O, A and B are collinear. For this, the associative law essentially governs the manner in which the definition must be made; that is to say, $OA + OB$ should be required to be a vector which makes equation (1) true for every choice of C not on OAB. To show that such a vector exists for any particular choice of C is quite easy, but to prove that it is always the same vector—independent of the choice of C—is a matter requiring some discussion.

Next, having established the validity of (1) in all possible cases, one can introduce integral multiples of a vector by repeated addition, i.e.

$$n.OA = OA + OA + \ldots + OA \quad (n \text{ terms}),$$

and discuss the problem of finding the aliquot parts of a vector, which requires some interesting geometry; and, finally, one can assign a meaning to $r.OA$ for an arbitrary rational number r (positive, negative or zero). A few details, then, are necessary to complete the proof that, *under addition relative to t, the vectors with initial point O (subject to the modifications mentioned in* (i) *and* (ii) *above) form a vector space over the field of rational numbers.* (Multiplication of vectors by arbitrary real numbers may also be discussed.)

If one wishes to go on to consider addition of vectors with different initial points, it is necessary first to introduce the idea of *equivalent* vectors thus: two non-collinear vectors AB and A_1B_1 are equivalent (relative to t) if the lines AB

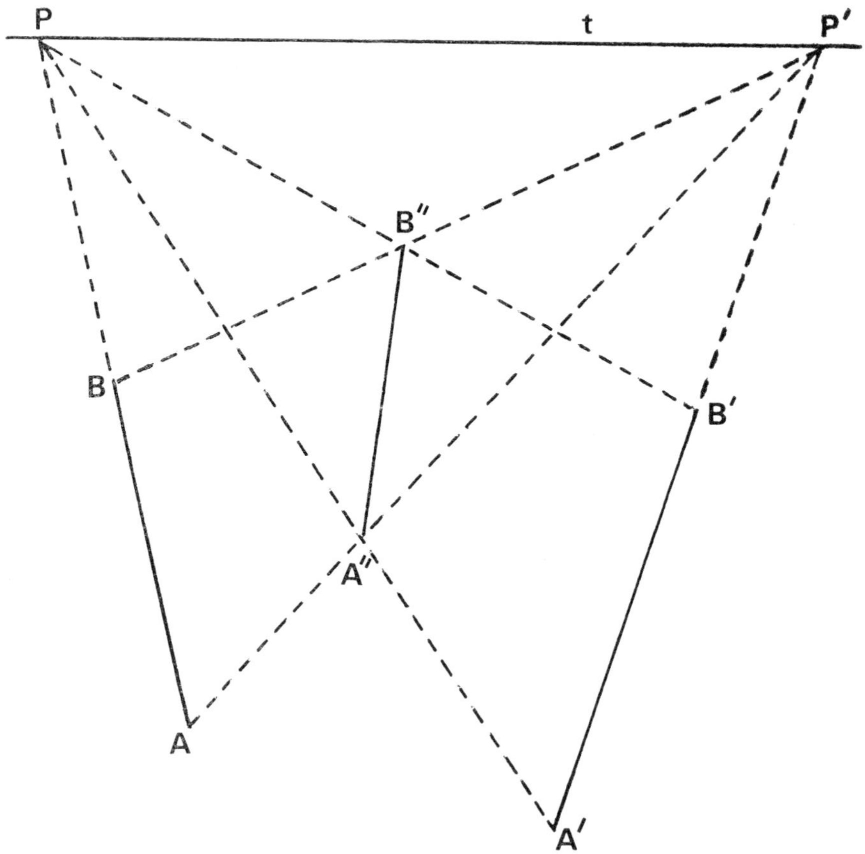

Figure 3 (see Section 1.1)

and A_1B_1 meet on t and if also the lines AA_1 and BB_1 meet on t. We express this relation by writing $AB \sim A_1B_1$. The proof that this is an equivalence relation (including the problem of deciding when two collinear vectors should be regarded as equivalent) is again an interesting exercise in geometry. Of course, if t is at infinity, two vectors are equivalent if they are equal in magnitude and direction (though they may have different initial points).

Returning then to the problem of defining the sum $AB + A'B'$ (when A and A' are different), the idea is to choose an arbitrary point O and to construct the two vectors OC and OC' with initial point O which are equivalent to AB and $A'B'$ respectively; OC and OC' may then be added according to the rule previously given, yielding a sum-vector OD. By choosing a different position for O we would, of course, obtain a different sum-vector OD, but it can be shown that any two of the possible sum-vectors OD are equivalent. It is therefore legitimate to write

$$AB + A'B' \sim OD$$

with a sign of equivalence.

Various topics may then be investigated, but we content ourselves here with mention of just one more result, namely: *if (as in Figure* 3) *AB and $A'B'$ are two coplanar vectors whose lines meet t in P and P', and if A'' and B'' are the points of intersection of the lines AP' and $A'P$ and of the lines BP' and $B'P$ respectively, then*

$$AB + A'B' \sim A''B''.$$

This result gives an elegant construction for the sum, up to equivalence, of two coplanar vectors with different initial points.† By taking the end-point of AB to coincide with the initial point of $A'B'$, we obtain the following noteworthy special case: *if A, B, C are three arbitrary points not on t, then*

$$AB + BC \sim AC.$$

This shows that a well-known result for ordinary vector addition carries over, without essential change, to the more general theory of projective addition.

1.2. LINE-LINE VECTORS

As we have already said in Section 1.1, a vector is really nothing more than an ordered pair of points. Therefore, by dualising the theory of point-point vectors, it is possible to develop a corresponding theory of line-line vectors in the plane (or of plane-plane vectors in space). Thus a *line-line vector* is just an ordered pair of lines ab (a being the *initial line* and b the *end-line*) and one can introduce projective addition of line-line vectors relative to a point T. This is done, in the first instance, for line-line vectors which share the same initial line, the sum oc of two line-line vectors oa and ob being defined by the construction shown in Figure 4 (which is dual to Figure 1a). The theory then proceeds in exactly the same way as for point-point vectors (with the necessary dual changes of language).

†It will be noted that this construction is not itself associative; if we apply it to obtain the sum of three coplanar vectors, we get two different sum-vectors according to the method of bracketing chosen; but the two sum-vectors are, of course, equivalent. It will also be noted that the construction only makes sense for coplanar vectors and does not work for arbitrary vectors in space.

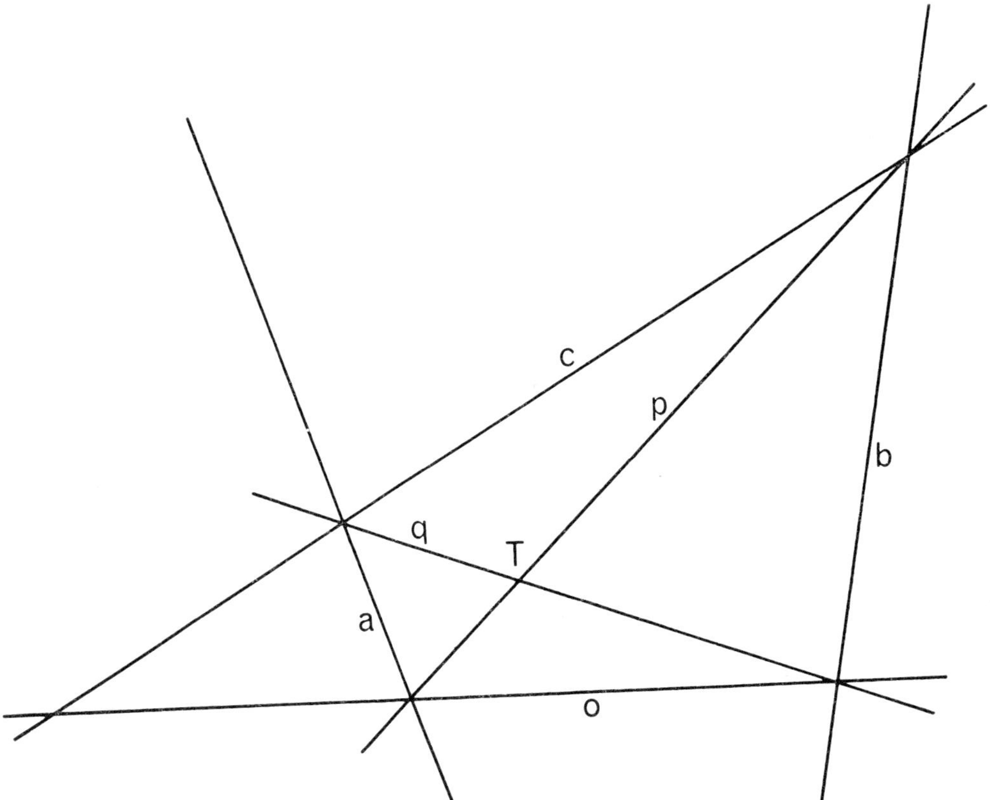

Figure 4 (see Section 1.2)

If one has worked carefully through the theory of point-point vectors, a repeat performance in terms of line-line vectors is rather tedious. The following, however, are a few details which are worth some attention.

(i) Just as for point-point vectors (cf. notes (i) and (ii) of Section 1.1), certain line-line vectors turn out to be 'exceptional' and have to be excluded from consideration in order to retain a well-defined operation of addition for line-line vectors; and this same reason compels the introduction of 'ideal' line-line vectors (for which one of the two lines is the line at infinity).

(ii) It is useful to examine the construction for addition of line-line vectors for the special case in which T is taken to be a point at infinity.

(iii) The diagram needed to illustrate the associative law

$$(oa + ob) + oc = oa + (ob + oc)$$

provides an interesting exercise in the drawing of dual figures (the required figure being the dual of Figure 2).

(iv) Two other useful exercises in duality are (a) to formulate the definition of equivalent line-line vectors and (b) to give a precise geometrical interpretation of the equivalence $ab + bc \sim ac$.

SECTION 2

EXTENSIONS OF PASCAL'S AND BRIANCHON'S THEOREMS

Pascal's theorem states that, if a hexagon is inscribed in a conic, then its three pairs of opposite sides intersect in three points which lie on a line (called the Pascal line of the hexagon). The dual result, Brianchon's theorem, states that, if a hexagon is circumscribed to a conic, then its diagonals are concurrent (at a point called the Brianchon point of the hexagon). The two theorems are illustrated, with dual letterings, in Figures 5 and 6 (where, in the former, we have followed the usual custom of taking the hexagon to be re-entrant, so as to get a compact figure).

Besides being of interest in themselves, these theorems are important as starting points for various lines of geometrical enquiry, and the object of this Section is to pursue some of these. In Section 2.1, we present what we believe to be a new theorem concerning the Pascal lines of the seven hexagons obtained from a heptagon inscribed in a conic by omitting each of its vertices in turn. Next, in Sections 2.2-2.4, we describe a series of theorems which generalize Pascal's theorem (in much the same way as Pascal's theorem itself generalizes the simpler theorem of Pappus). Finally, in Section 2.5, we describe an elegant theorem which concerns octagons, rather than hexagons, inscribed in a conic.

2.1. THE HEPTAGON THEOREM

Pascal's theorem is essentially about six points *taken in a given cyclic order* on a conic. Our first theorem concerns the Pascal hexagons derivable from a set of seven points taken in cyclic order on a conic. Thus:

The heptagon theorem. Let H be a heptagon (whose seven vertices are in a given cyclic order) inscribed in a conic. Then the Pascal lines of the seven hexagons obtained by omitting each vertex of H in turn (and keeping the remaining vertices in the same cyclic order) are the sides of a heptagon I which circumscribes a conic. Moreover, the Brianchon points of the seven hexagons obtained by omitting the sides of I one at a time (and keeping the remaining sides in the natural cyclic order) are the vertices of the original heptagon H.

Proof. We denote the vertices of the heptagon H by 1, 2, . . . , 7 in order (see Figure 7) and we introduce names for other points as follows:

$$B = (45.71),\ C = (56.12),\ D = (67.23),\ E = (71.34),$$
$$F = (12.45),\ X = (23.56),\ Y = (34.67),\ A = (FX.BY),$$

where $B = (45.71)$ indicates that B is the intersection of the line joining 4 to 5 with the line joining 7 to 1, and so on.

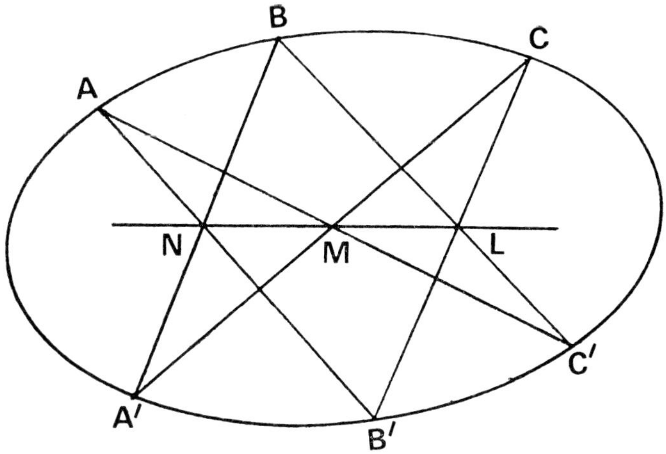

Figure 5 (see foreword to Section 2)

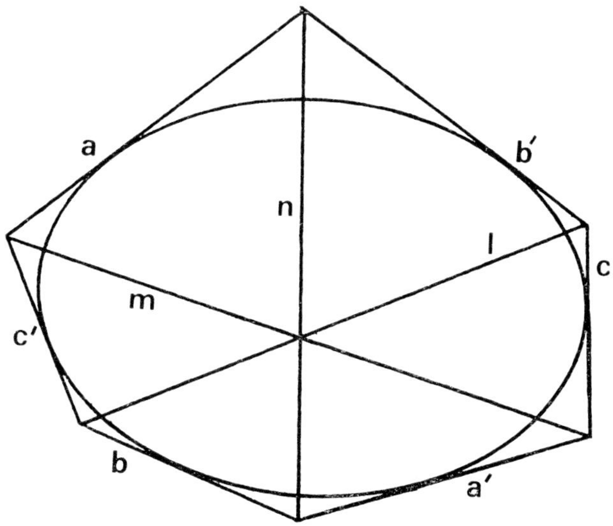

Figure 6 (see foreword to Section 2)

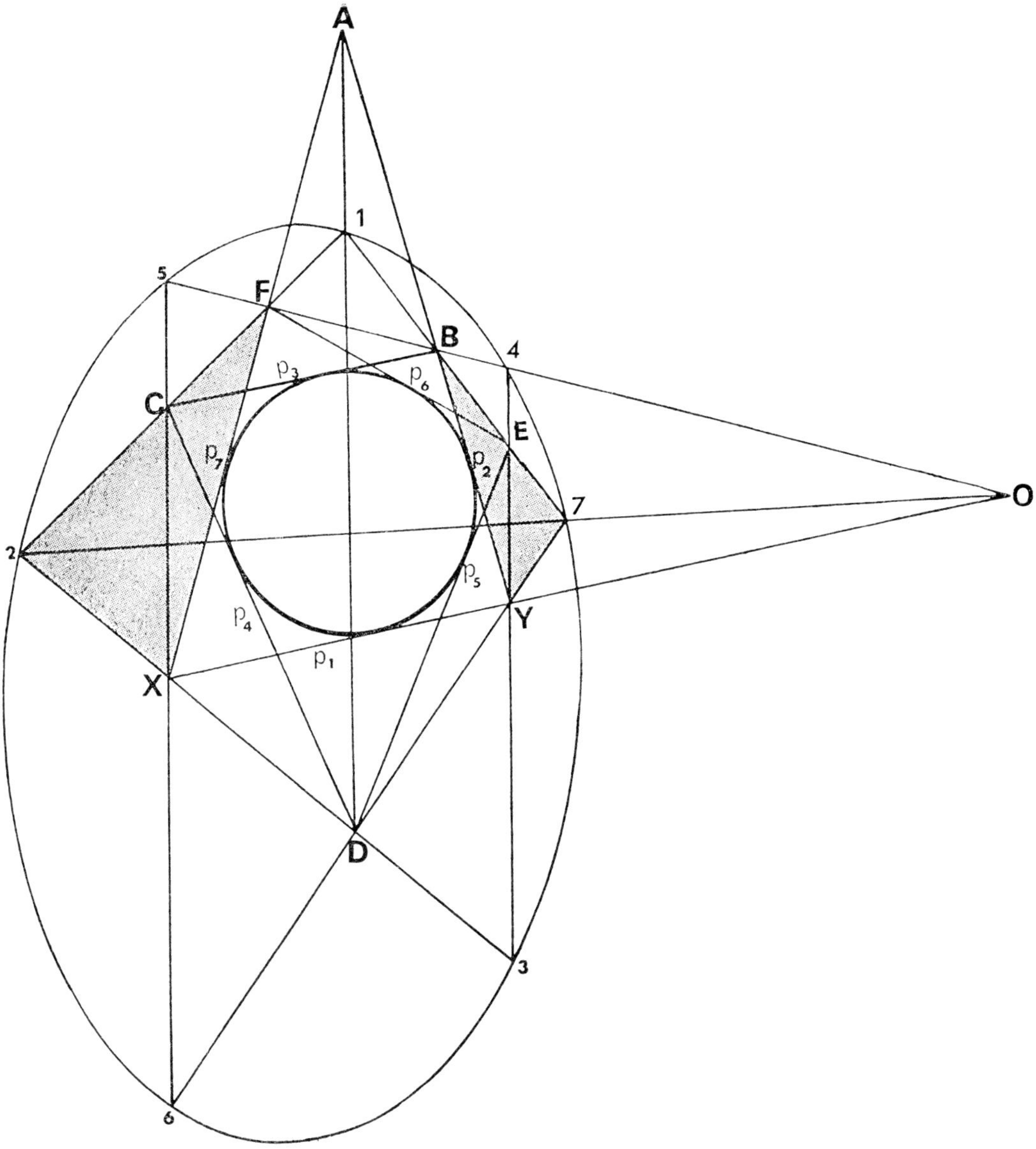

Figure 7 (see Section 2.1)

If, then, $H_1, H_2, \ldots, H_7$ denote the ordered hexagons obtained from H by omitting the vertices $1, 2, \ldots, 7$ in turn, and if $p_1, p_2, \ldots, p_7$ denote their respective Pascal lines, we easily identify these lines as follows:

$$p_1 = XY, \quad p_2 = YB, \quad p_3 = BC, \quad p_4 = CD,$$
$$p_5 = DE, \quad p_6 = EF, \quad p_7 = FX.$$

As regards H_1, we note that the third point on its Pascal line p_1 is (45.72); it follows that the lines 45, 72 and XY are concurrent, or (since F and B lie on the line 45) that the triangles $2FX$ and $7BY$ are in perspective. (These triangles are shown shaded in Figure 7, their vertex of perspective being denoted by O.) By Desargues's theorem, therefore, the corresponding sides of these triangles meet in collinear points, i.e. $(FX.BY)$, $(2X.7Y)$ and $(2F.7B)$ are collinear. Since these points are A, D and 1 respectively, it follows that AD passes through the point 1 (which is where BE meets CF). Hence, by the converse of Brianchon's theorem, $ABCDEF$ is a Brianchon hexagon with Brianchon point 1; and, since the sides AB, BC, CD, DE, EF, FA of this hexagon are respectively $p_2, p_3, \ldots, p_7$, this is the same as saying that $p_2, p_3, \ldots, p_7$ are the sides of a Brianchon hexagon with 1 as Brianchon point. In particular, p_2 touches the (unique) conic k which can be drawn to touch $p_3, p_4, \ldots, p_7$.

In like manner (or just by cyclic symmetry), we find that $p_1, p_3, \ldots, p_7$ are the sides of a Brianchon hexagon (with 2 as Brianchon point) and hence that p_1 also touches the conic k which touches $p_3, p_4, \ldots, p_7$. The conic k is therefore touched by all seven lines $p_1, p_2, \ldots, p_7$ and the first clause of our theorem is established; moreover, since 1 is the Brianchon point of $p_2, \ldots, p_7$, and so on cyclically, the second clause is also established.

It might be of interest to investigate the complete figure obtained from seven points of a conic by considering them in the 360 different cyclic orders that are possible; this figure would comprise 420 Pascal lines and 360 conics, each line touching six of the conics and each conic touching seven of the lines. We shall not, however, pursue this matter here. (A corresponding investigation of the figure of 60 Pascal lines associated with six points of a conic may be found in the Appendix to Salmon's 'Treatise on Conic Sections', 6th edition, London, 1879.)

2.2. THE THREE-CONICS THEOREM

It is a common device, in projective geometry, to take a theorem which involves a proper conic k, to allow k to degenerate into a pair of lines, and thereby to obtain another theorem as a special case of the first. The theorem of Pappus, for example, can be obtained in this way as a special case of Pascal's theorem. We propose now to describe an even more general theorem from which Pascal's theorem can be obtained as a special case. We call it the three-conics theorem and state it thus:

The three-conics theorem. If three conics pass through two given points, then the lines joining the other two intersections of each pair of the conics are concurrent.

It will be noted that, if two of the conics become line-pairs, the theorem reduces to that of Pascal, while, if all three conics are line-pairs, we obtain

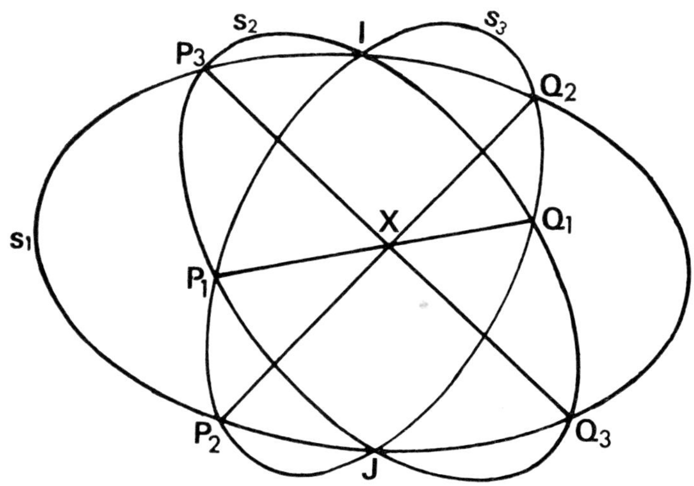

Figure 8 (see Section 2.2)

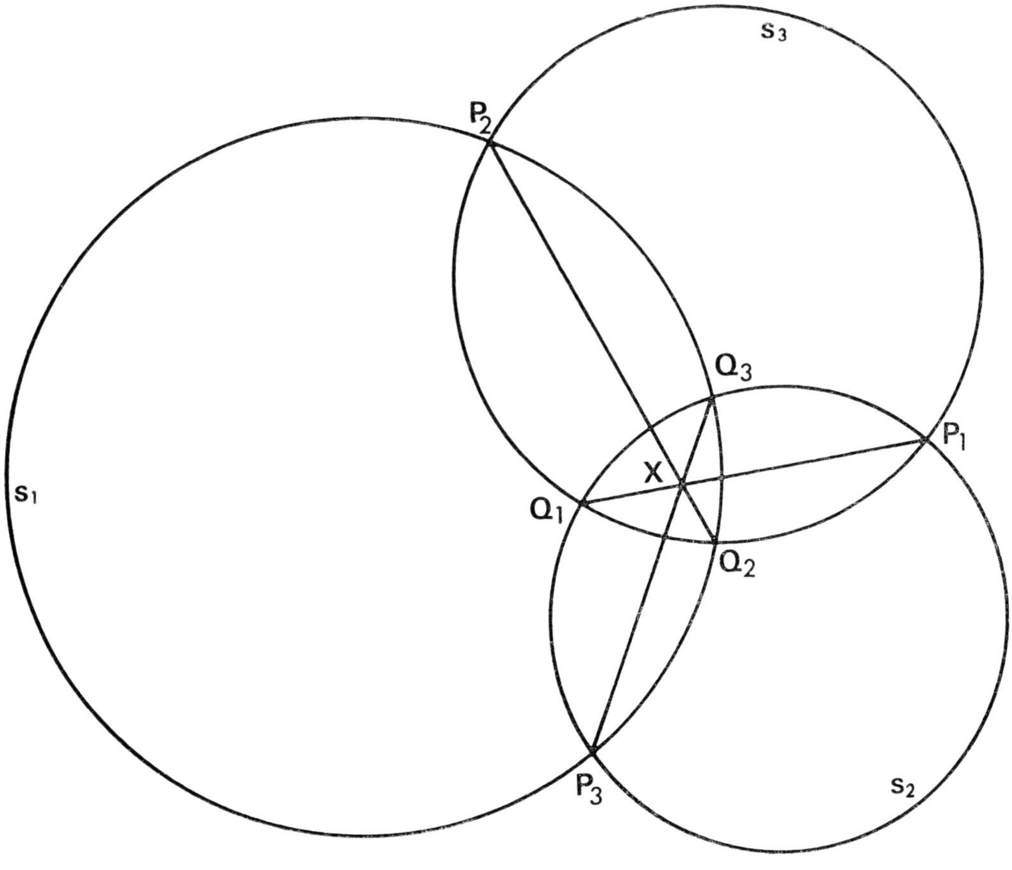

Figure 9 (see Section 2.2)

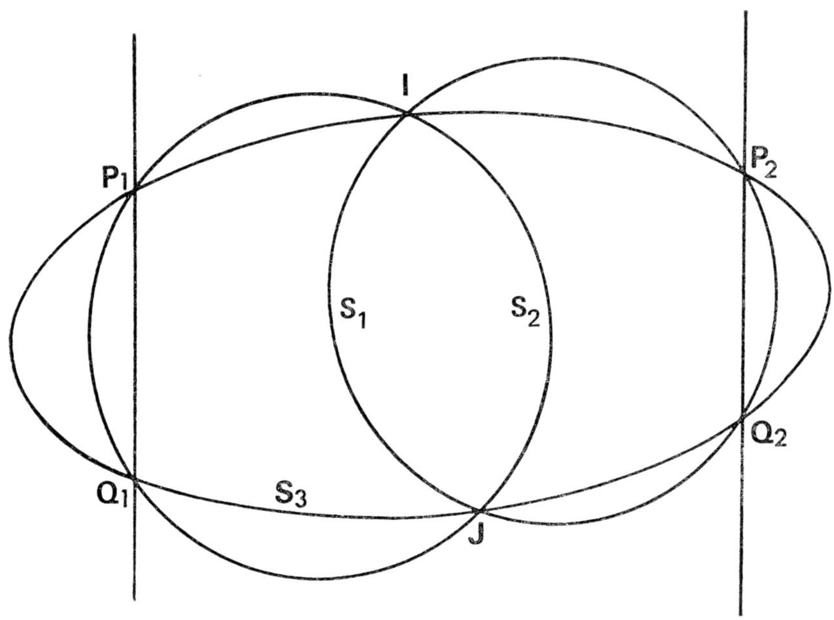

Figure 10 (see Section 2.2)

Pappus's theorem. There is also another special case, intermediate between the general theorem and Pascal's theorem, obtained by allowing just one of the three conics to degenerate.

Proof. Figure 8 illustrates the theorem, with three conics S_1, S_2, S_3 passing through two points I and J, the remaining intersections of S_2 and S_3 being P_1 and Q_1, while those of S_3 and S_1 are P_2 and Q_2, and those of S_1 and S_2 are P_3 and Q_3. The theorem asserts that P_1Q_1, P_2Q_2 and P_3Q_3 are concurrent.

To see this, let the equation of S_1, in a chosen co-ordinate system, be $S_1 = 0$ and let the equations of the lines IJ, P_2Q_2 and P_3Q_3 be $L = 0$, $L_2 = 0$ and $L_3 = 0$ respectively. Then S_2, being a member of the pencil of conics determined by S_1 and the line-pair (IJ, P_3Q_3), will have an equation of the form

$$S_1 + \lambda LL_3 = 0$$

where λ is a constant; and, similarly, S_3 will have an equation of the form

$$S_1 + \mu LL_2 = 0$$

where μ is a constant. The equation

$$(S_1 + \lambda LL_3) - (S_1 + \mu LL_2) \equiv L\,(\lambda L_3 - \mu L_2) = 0$$

therefore represents a conic k of the pencil determined by S_2 and S_3, i.e. a conic through I, J, P_1 and Q_1. Since this conic k consists of the two lines $L = 0$ and $\lambda L_3 - \mu L_2 = 0$, of which the former is IJ, it follows that the latter must be P_1Q_1. We have thus shown that the equation of P_1Q_1 is $\lambda L_3 - \mu L_2 = 0$, which is a linear combination of the equations of P_2Q_2 and P_3Q_3. It follows that P_1Q_1 belongs to the pencil of lines determined by P_2Q_2 and P_3Q_3 and the three-conics theorem is thereby established.

A less elementary proof of the same theorem may be given by making use of the theorem of *nine associated points*, namely, that *any cubic curve that passes through eight of the nine intersections of two given cubic curves automatically passes through their ninth intersection.* (A proof of this may be found in any treatise on higher plane curves.) Thus, taking the given cubics to be the reducible curves consisting of S_2 and P_2Q_2 and of S_3 and P_3Q_3, we see that the reducible cubic consisting of S_1 and P_1Q_1 passes through eight of their nine intersections (namely I, J, P_1, Q_1, P_2, Q_2, P_3, Q_3) and so passes through the ninth (which is where P_2Q_2 meets P_3Q_3). As this ninth point plainly does not lie on S_1, it must lie on P_1Q_1 and we have that P_1Q_1, P_2Q_2 and P_3Q_3 are concurrent at this ninth point (denoted by X in Figure 8).

Two Euclidean versions of the theorem are perhaps worth mentioning:

(i) If I and J are taken to be the circular points at infinity, we obtain the familiar theorem (see Figure 9) that *the radical axes of three circles, taken by pairs, are concurrent.*

(ii) If, on the other hand, we take P_3 and Q_3 to be the circular points at infinity, we obtain the less familiar result:

If two circles S_1 and S_2 pass through two points I and J on a conic S_3, then the lines P_1Q_1 and P_2Q_2 joining their remaining points of intersection with S_3 are parallel. This version is illustrated in Figure 10.

The converse of the three-conics theorem may be stated as follows:

If two conics S_2 and S_3 meet at four points I, J, P_1, Q_1, and if P_2Q_2 and

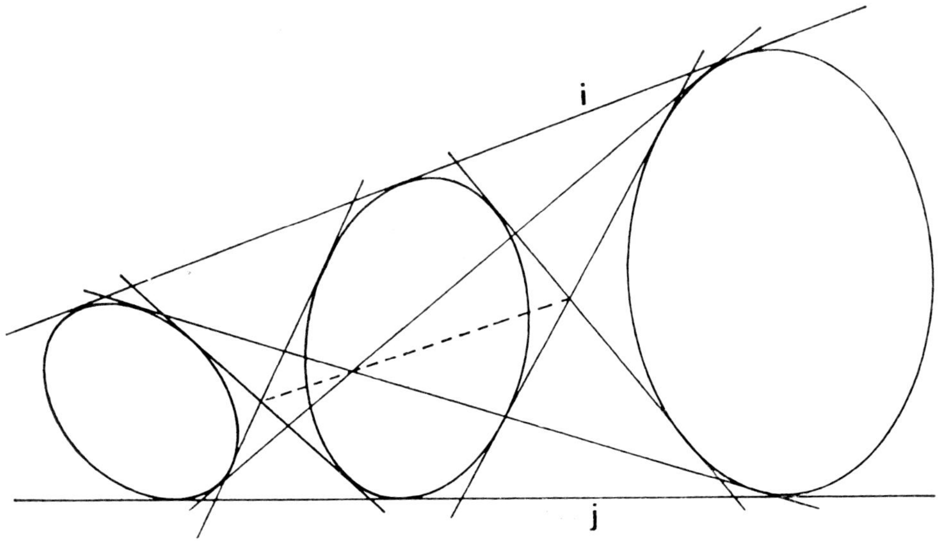

Figure 11 (see Section 2.2)

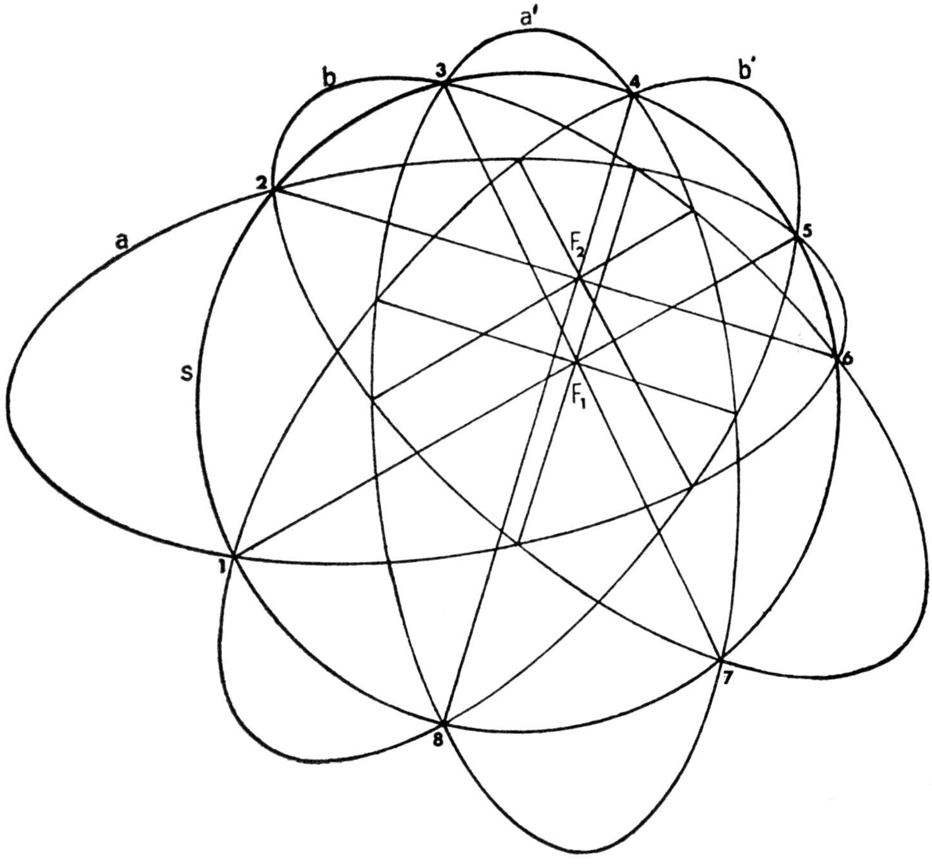

Figure 12 (see Section 2.2)

P_3Q_3 are chords of S_3 and S_2 respectively which meet on P_1Q_1, then the six points I, J, P_2, Q_2, P_3, Q_3 lie on a conic.

The proof is by a simple *reductio ad absurdum* argument.

The dual of the three-conics theorem is also of interest. This is illustrated in Figure 11 and states:

If three conics share two common tangents i and j, then their remaining pairs of common tangents intersect at three collinear points.

An amusing compound of the three-conics theorem is shown in Figure 12. Here we have eight points 1, 2, . . . , 8 on a conic S and four conics a, b, a', b' are drawn through 1256, 2367, 3478 and 4581 respectively. Then, by applying the three-conics theorem to each of the triads of conics (S, a, b), (S, b, a'), (S, a', b') and (S, b', a), one easily discovers two concurrent tetrads of lines (seen passing through F_1 and F_2 in the figure).

2.3. THE DOUBLE-CONTACT THEOREM

We have already seen, in the last section, that if three conics have two points in common, then their 'other' common chords (taken by pairs) are concurrent. If, however, we begin with three conics which (by pairs) have concurrent common chords, it is obviously not true that they must necessarily share two common points. The correct theorem in this connexion is as follows:

The double-contact theorem. If S_1, S_2 and S_3 are three conics having the property that there is a point X (not on any of the conics) lying on a common chord of each pair of the three conics (the chords in question being distinct), then there exists a conic ω which has double contact with each of S_1, S_2 and S_3.

Proof. Since the three given common chords are distinct and concurrent, we can take their equations (in a chosen co-ordinate system) as $L_1 = 0$, $L_2 = 0$ and $L_3 = 0$, where

$$L_1 + L_2 + L_3 \equiv 0. \tag{1}$$

We suppose that $L_1 = 0$ is a common chord of S_2 and S_3, that $L_2 = 0$ is a common chord of S_3 and S_1, and that $L_3 = 0$ is a common chord of S_1 and S_2. If then $S_3 = 0$ is the equation of S_3, the equation of S_2 will be of the form $S_3 + \lambda L_1 M_1 = 0$, where $M_1 = 0$ is the common chord of S_2 and S_3 opposite to $L_1 = 0$. It is no loss of generality, in fact, to take $\lambda = 1$ (since this factor can be absorbed into M_1). The equation of S_2 is then $S_2 = 0$, where

$$S_2 - S_3 \equiv L_1 M_1. \tag{2}$$

Similarly, the equation of S_1 can be taken as $S_1 = 0$, where

$$S_3 - S_1 \equiv L_2 M_2, \tag{3}$$

$M_2 = 0$ being an equation for the common chord of S_3 and S_1 opposite to $L_2 = 0$.

If we eliminate S_3 between the identities (2) and (3), we obtain

$$S_1 - S_2 \equiv -L_1 M_1 - L_2 M_2. \tag{4}$$

Now the point X (which lies on $L_1 = 0$ and $L_2 = 0$) must, by (4), satisfy $S_1 - S_2 = 0$; but there is only one conic of the pencil determined by S_1 and

S_2 which passes through X and (since X is on the common chord $L_3 = 0$ of S_1 and S_2) this conic consists of $L_3 = 0$ and the opposite common chord of S_1 and S_2, say $M_3 = 0$. We must therefore have

$$S_1 - S_2 \equiv L_3 M_3. \tag{5}$$

Now, from (4) and (5), we obtain the identity

$$L_1 M_1 + L_2 M_2 + L_3 M_3 \equiv 0$$

and, if we use (1) to eliminate L_3, this reduces to

$$L_2 (M_2 - M_3) \equiv L_1 (M_3 - M_1).$$

Since L_1, L_2 are distinct (non-proportional) linear expressions, the last identity shows that there must exist a constant k such that

$$M_2 - M_3 \equiv kL_1 \tag{6}$$

$$\text{and} \quad M_3 - M_1 \equiv kL_2 \tag{7}$$

and these (by use of (1)) imply a third similar relation

$$M_1 - M_2 \equiv kL_3. \tag{8}$$

At this point, we may notice that equation (6) demonstrates the concurrence of $L_1 = 0$, $M_2 = 0$ and $M_3 = 0$, which are common chords of pairs of our three given conics—not the same three as were originally supposed to be concurrent; and (7) and (8) have similar interpretations. We have therefore incidentally proved the following:

If three conics are such that, taken by pairs, they have three concurrent common chords, then they actually possess this property in four separate ways.

Returning, however, to the proof of the double-contact theorem, we now notice the identity

$$4kS_1+(M_2+M_3-M_1)^2 \equiv 4kS_2+(M_3+M_1-M_2)^2 \equiv 4kS_3+(M_1+M_2-M_3)^2. \tag{9}$$

The equality of the first two terms, for example, is equivalent to

$$4k(S_1-S_2) \equiv (M_3+M_1-M_2)^2-(M_2+M_3-M_1)^2 \equiv 4M_3(M_1-M_2)$$

which is a simple consequence of (5) and (8); and the second equality is similarly proved. If, therefore, we use Ω to denote any one of the three identical expressions (9), we see that the conic $\Omega = 0$ has double contact with each of $S_1 = 0$, $S_2 = 0$ and $S_3 = 0$ (the chords of contact being $M_2 + M_3 - M_1 = 0$, $M_3 + M_1 - M_2 = 0$ and $M_1 + M_2 - M_3 = 0$ respectively). The double-contact theorem is therefore proved, $\Omega = 0$ being the equation of the desired conic ω.

The converse is rather easier to establish:

If three conics S_1, S_2, S_3 all have double contact with another conic ω, then each two of S_1, S_2 and S_3 have a 'distinguished' pair of opposite common chords, the three such pairs of common chords being the pairs of opposite sides of a complete quadrangle.

It is possible, of course, to select a pair of opposite common chords of S_1 and S_2, say, in three different ways and, for the result just stated to be true, it is necessary to choose the 'right' pair. This explains our use of the word 'distinguished' in the above statement; the way of distinguishing the 'right' pair from the other two pairs will be made clear in the proof below.

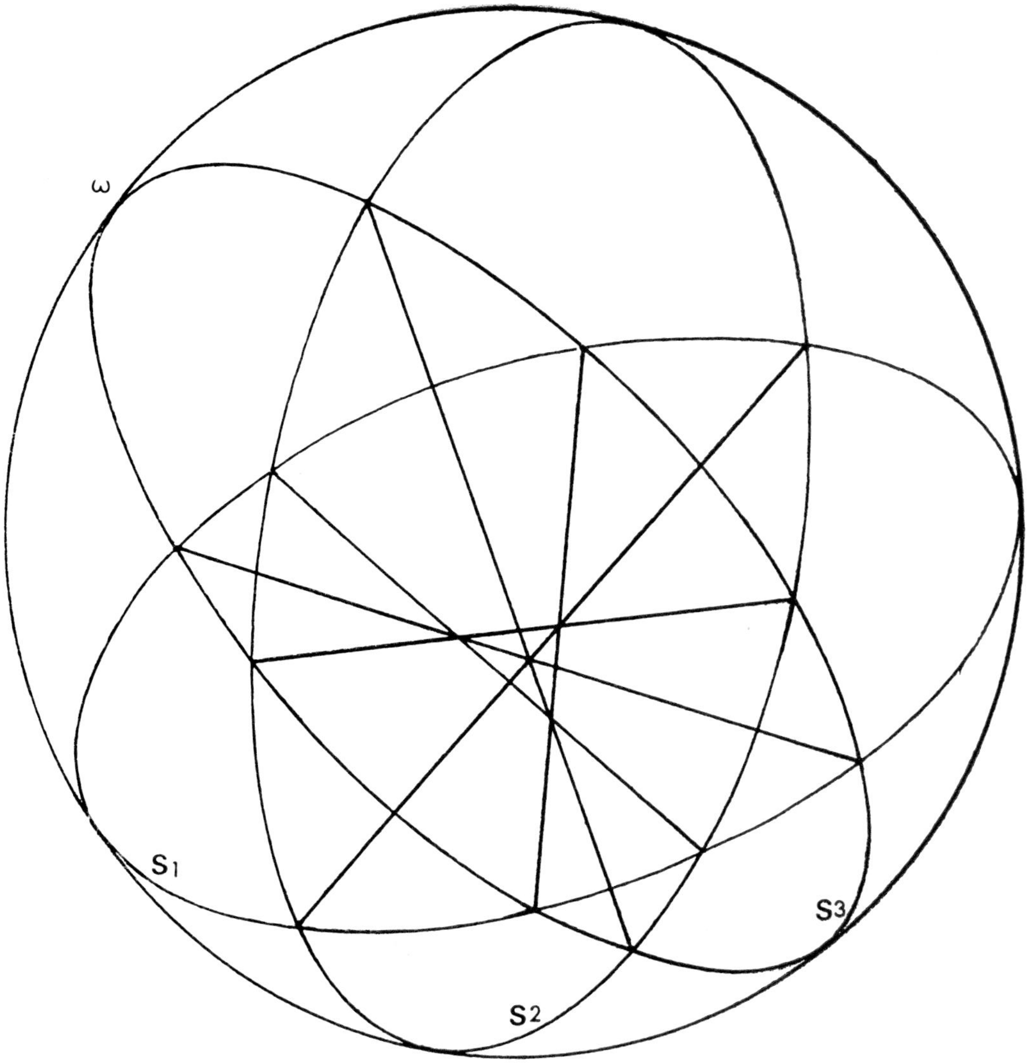

Figure 13 (see Section 2.3)

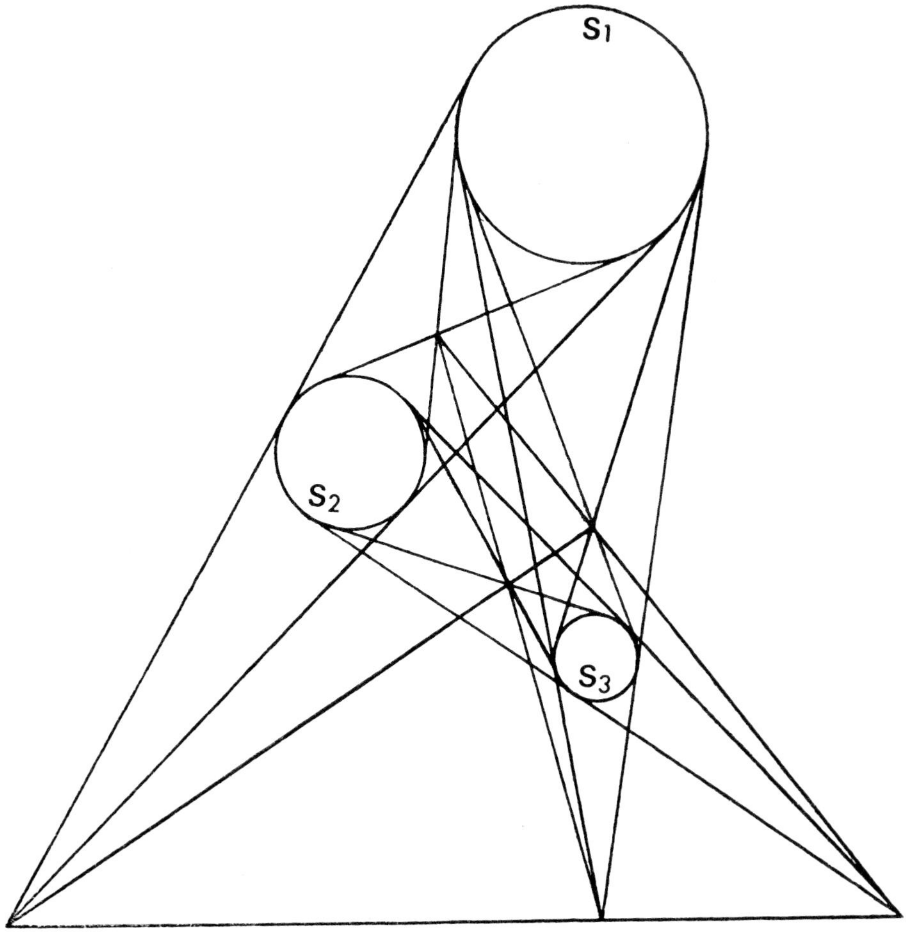

Figure 14 (see Section 2.3)

Proof. If we take the equation of ω to be $\Omega = 0$, then the equations of S_1, S_2, S_3 will take the respective forms

$$\Omega + N_1^2 = 0, \quad \Omega + N_2^2 = 0, \quad \Omega + N_3^2 = 0,$$

where $N_1 = 0$, $N_2 = 0$ and $N_3 = 0$ are the three chords of contact with ω. We notice then that

$$(\Omega + N_2^2) - (\Omega + N_3^2) \equiv (N_2 + N_3)(N_2 - N_3) = 0$$

is the equation of a line-pair belonging to the pencil determined by S_2 and S_3, so that

$$N_2 + N_3 = 0 \quad \text{and} \quad N_2 - N_3 = 0 \tag{10}$$

are the equations of a pair of opposite common chords of S_2 and S_3. The form of these equations, moreover, shows that *these common chords pass through the point of intersection of the two chords of contact of S_2 and S_3 with ω (and, indeed, they harmonically separate these chords of contact); this is the property by which this particular pair of common chords is 'distinguished' from the other two pairs.* Similarly, the equations of the distinguished pairs of common chords of S_3 and S_1 and of S_1 and S_2 are

$$N_3 + N_1 = 0, N_3 - N_1 = 0 \quad \text{and} \quad N_1 + N_2 = 0, N_1 - N_2 = 0 \tag{11}$$

respectively; and, as is easily seen, the six lines given by (10) and (11) are the sides of the quadrangle whose vertices are given by

$$N_1 = \pm N_2 = \pm N_3$$

for each of the four different sign combinations. This proves the converse of the double-contact theorem.

Figure 13 illustrates the configuration of the double-contact theorem and its converse. (It will be noted that, when ω degenerates (as an envelope) into a point-pair, the converse of the double-contact theorem reduces precisely to the three-conics theorem of the preceding section.) As regards the dual theorems, these are easily written down, as follows:

If three conics are such that (taken by pairs) they have couples of common tangents intersecting at three distinct points on a line (not itself a tangent to any of the conics) then (a) the conics have this property in four different ways and (b) the conics all have double contact with a fourth. And, conversely, if three conics each have double contact with a fourth, then certain of their common tangents intersect by pairs at the vertices of a complete quadrilateral.

We have not included a figure to illustrate this dual theorem in its full generality but we have (Figure 14) drawn a diagram to illustrate the special case in which the conic with which the other three have double contact reduces (as an envelope) to the pair of circular points at infinity—the statement of this special case being:

The six centres of similitude of three circles (taken by pairs) are the vertices of a complete quadrilateral.

2.4. THE FOUR-CONICS THEOREM

One other aspect of the figure described in the previous section is worthy of note. Thus, further to the proof of the double-contact theorem, we notice

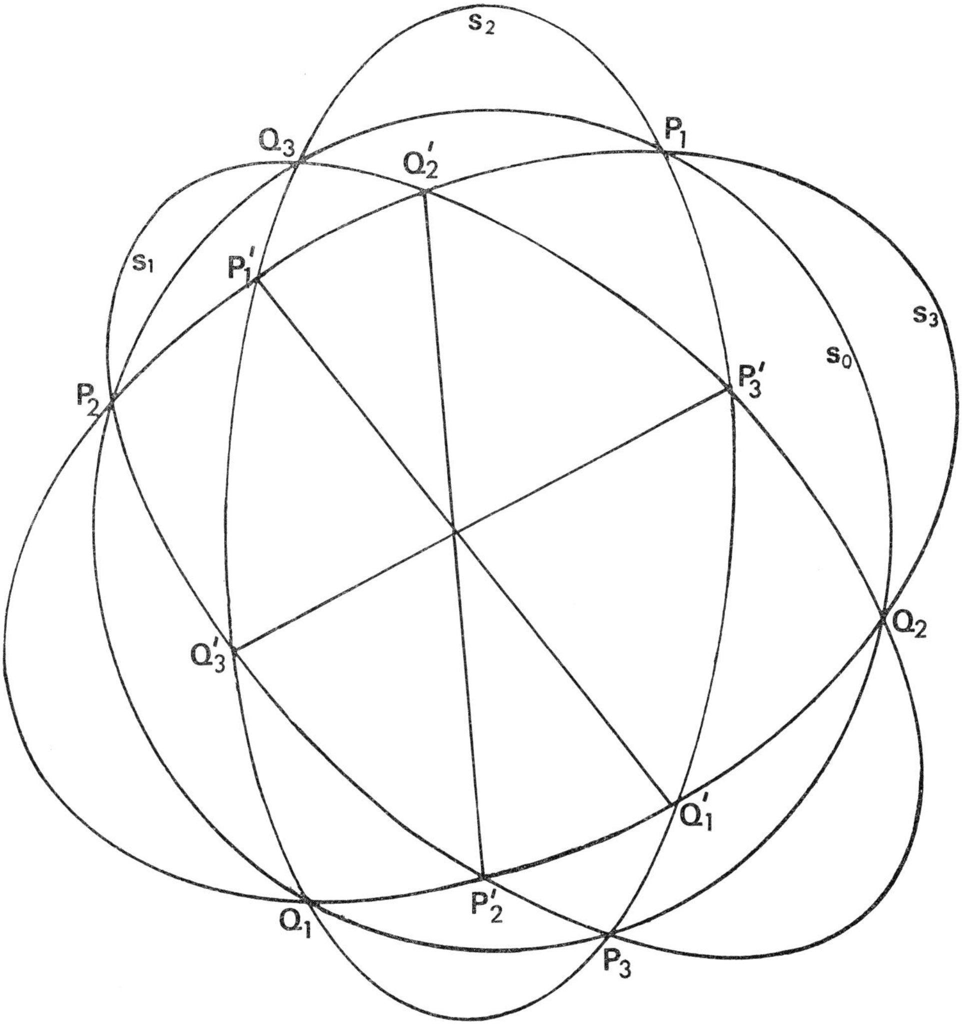

Figure 15 (see Section 2.4)

C

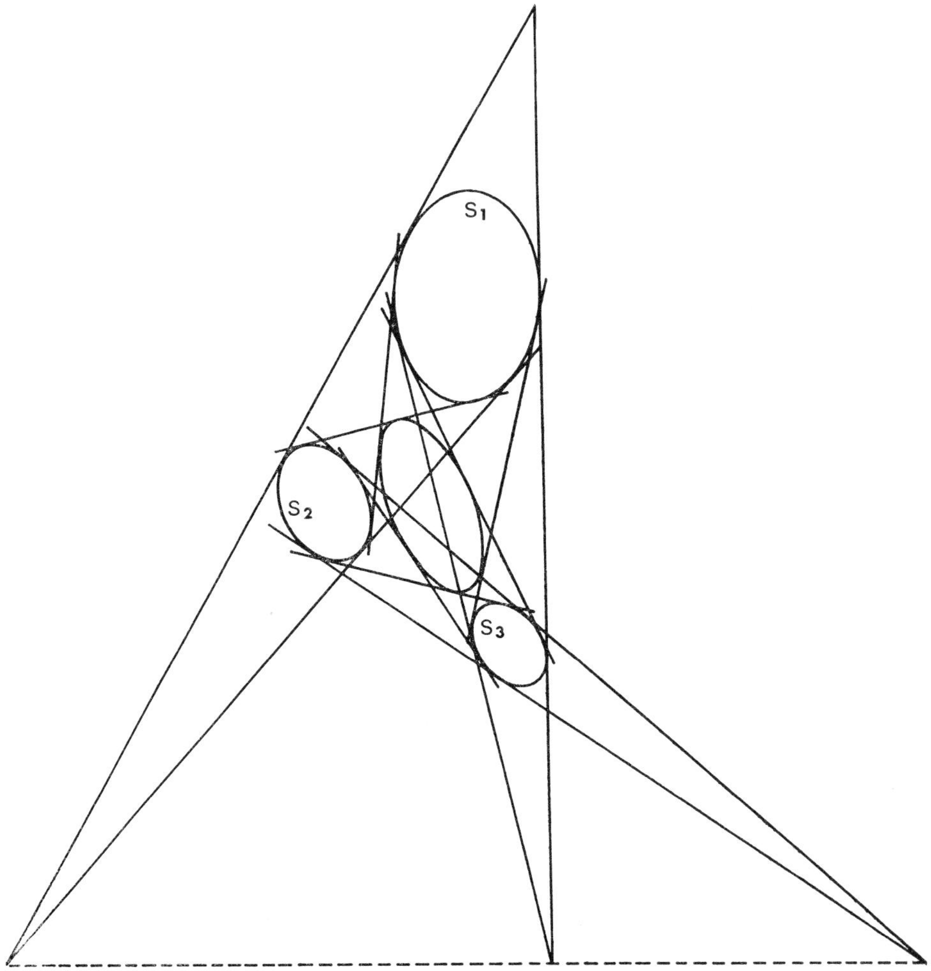

Figure 16 (see Section 2.4)

another identity, namely

$$kS_1 + M_2M_3 \equiv kS_2 + M_3M_1 \equiv kS_3 + M_1M_2. \tag{1}$$

To see this, observe that the first identity here is equivalent to $k(S_1 - S_2) \equiv M_3(M_1 - M_2)$, which is a simple consequence of equations (5) and (8) of Section 2.3; and the second identity is similarly proved. If, therefore, we let S_0 stand for either of the identical expressions (1), we see from $S_0 \equiv kS_1 + M_2M_3$ that the conic $S_0 = 0$ passes through the intersections of $S_1 = 0$ with $M_2 = 0$ and $M_3 = 0$; and the other two forms of S_0 have similar interpretations. We have thus proved the following theorem:

If three conics satisfy the hypothesis of the double-contact theorem, so that six of their twelve points of intersection by pairs determine three concurrent lines, then the other six lie on a conic.

The twelve points thus split up into two sets of six, one set forming the vertices of a Brianchon hexagon, while the other set forms a Pascal hexagon. Moreover (by the result proved incidentally in the course of the proof of the double-contact theorem) the twelve points can be so split up in four essentially different ways.

The converse of the last theorem is particularly simple and we call it the four-conics theorem. It may be stated thus:

The four-conics theorem. If two intersections of each pair of three conics S_1, S_2, S_3 lie on a conic S_0, then the lines joining the other two intersections of each pair are concurrent.

Proof. The theorem is illustrated in Figure 15 and we shall use the notation given there. We let the equation of S_0 be $S_0 = 0$ and we let the equations of P_1Q_1, P_2Q_2 and P_3Q_3 be $L_1 = 0$, $L_2 = 0$ and $L_3 = 0$ respectively. Then S_1, being a member of the pencil determined by S_0 and the line-pair (P_2Q_2, P_3Q_3), will have an equation of the form $S_1 = 0$, where $S_1 \equiv \lambda_1 S_0 + L_2L_3$, λ_1 being a constant. Similarly S_2 and S_3 will have equations $S_2 = 0$ and $S_3 = 0$, where

$$S_2 \equiv \lambda_2 S_0 + L_3L_1 \quad \text{and} \quad S_3 \equiv \lambda_3 S_0 + L_1L_2,$$

for suitable constants λ_2, λ_3. Now

$$\lambda_2 S_3 - \lambda_3 S_2 \equiv L_1(\lambda_2 L_2 - \lambda_3 L_3)$$

and this shows that the line-pair consisting of $L_1 = 0$ and $\lambda_2 L_2 - \lambda_3 L_3 = 0$ belongs to the pencil determined by S_2 and S_3. But there is only one line-pair of this pencil having $L_1 = 0$ as a component, namely $(P_1Q_1, P_1'Q_1')$. It follows that the equation of $P_1'Q_1'$ is $\lambda_2 L_2 - \lambda_3 L_3 = 0$. Similarly, the equations of $P_2'Q_2'$ and $P_3'Q_3'$ are respectively $\lambda_3 L_3 - \lambda_1 L_1 = 0$ and $\lambda_1 L_1 - \lambda_2 L_2 = 0$. And now, since

$$(\lambda_2 L_2 - \lambda_3 L_3) + (\lambda_3 L_3 - \lambda_1 L_1) + (\lambda_1 L_1 - \lambda_2 L_2) \equiv 0,$$

the equation of $P_1'Q_1'$ is a linear combination of those of $P_2'Q_2'$ and $P_3'Q_3'$; and we deduce that $P_1'Q_1'$, $P_2'Q_2'$ and $P_3'Q_3'$ are concurrent, as required.

The dual of the four-conics theorem is illustrated in Figure 16. It states that:

If two common tangents of each pair of three conics touch a fourth conic, then the remaining common tangents of each pair intersect in three collinear points.

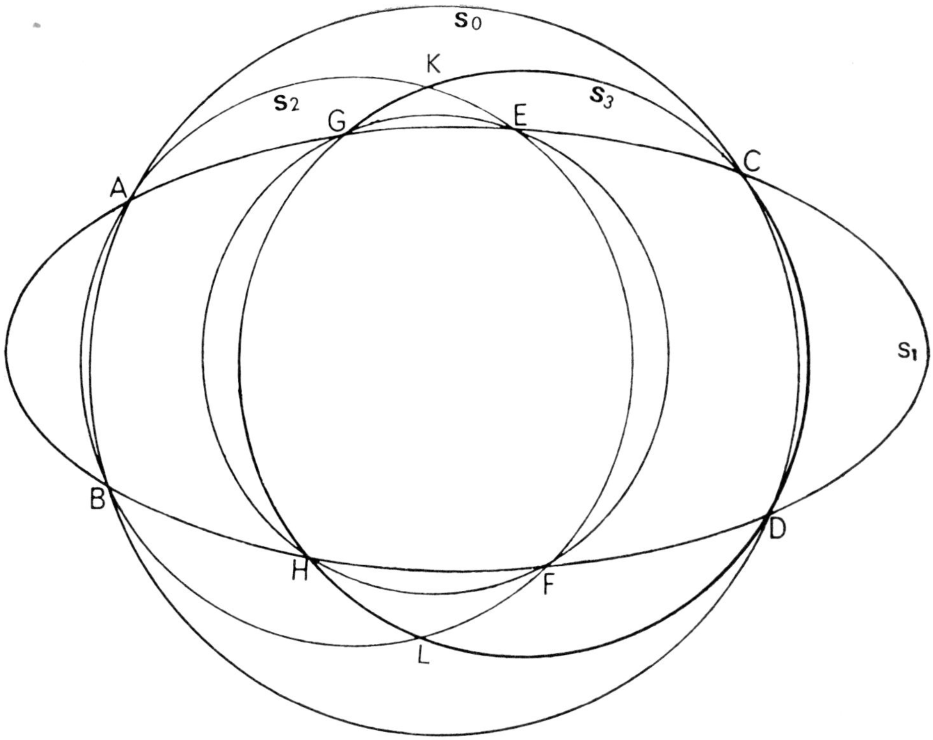

Figure 17 (see Section 2.4)

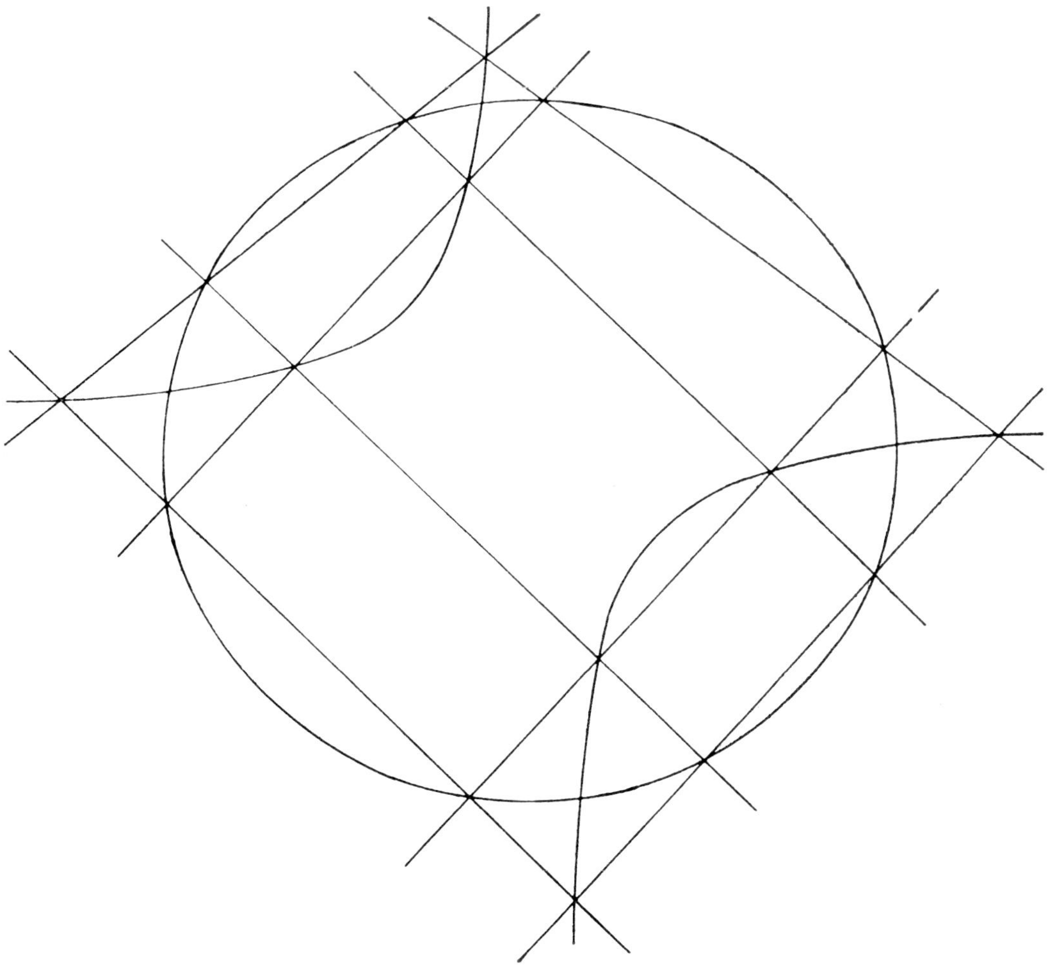

Figure 18 (see Section 2.5)

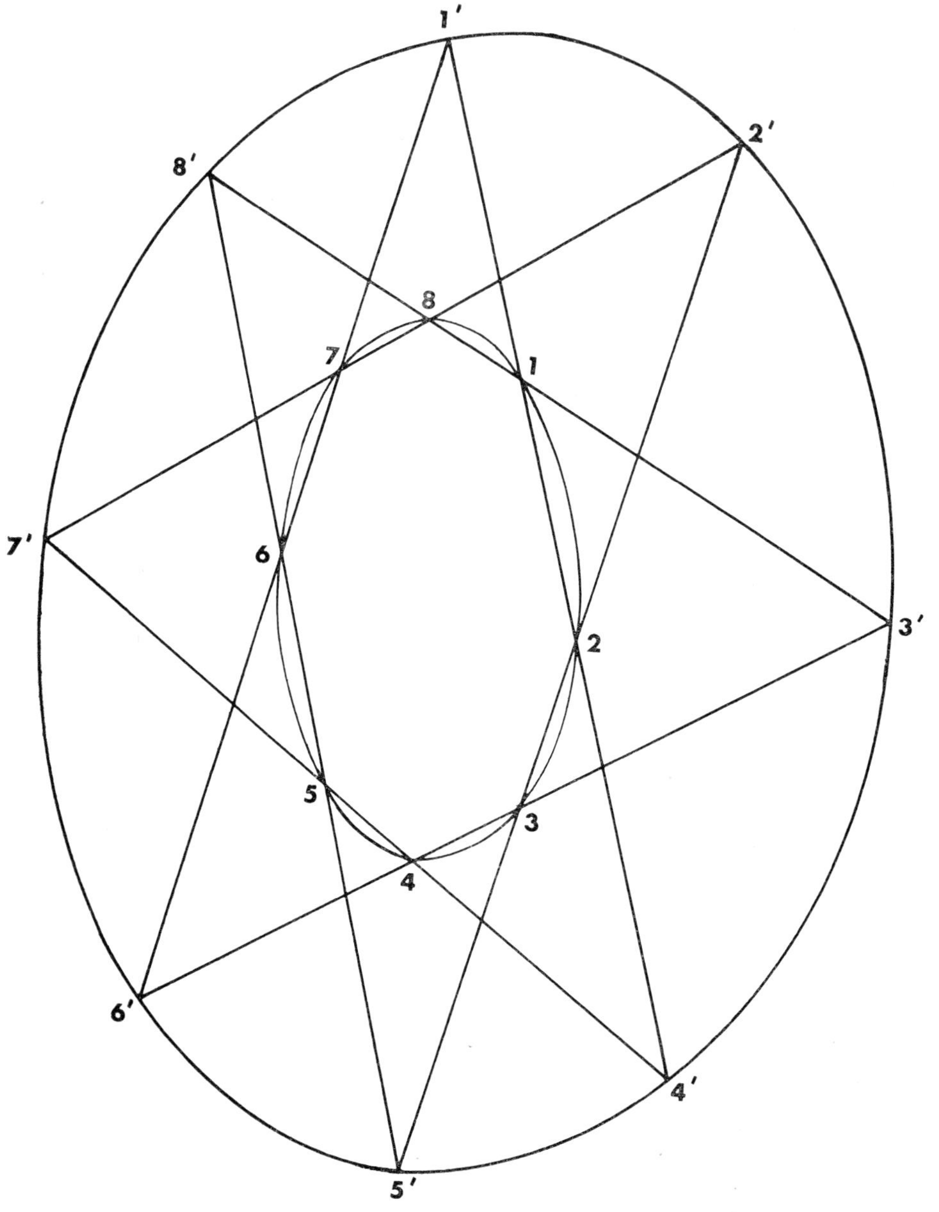

Figure 19 (See Section 2.5)

An interesting exercise in the application of the four-conics theorem and of the converse of the three-conics theorem (see Section 2.2) is illustrated in Figure 17. This shows a circle S_0 and a conic S_1 meeting at four points A, B, C, D and two other circles S_2 and S_3, of which the former passes through A and B and meets S_1 again at E and F, while the latter passes through C and D and meets S_1 again at G and H. It is then the case that E, F, G and H are concyclic. This may be proved by first applying the four-conics theorem to S_1, S_2, S_3 (each pair of which have two intersections on S_0) with the conclusion that EF and GH meet on the radical axis KL of S_2 and S_3; then, by the converse of the three-conics theorem (applied to S_2 and S_3) it follows that E, F, G and H lie on a conic through the 'other' two common points of S_2 and S_3 (besides K and L), which are of course the circular points at infinity; in other words, E, F, G and H are concyclic, as required.

2.5. INSCRIBED OCTAGONS

The last aspect of Pascal's theorem to which we wish to refer is its membership of a certain sequence of theorems concerning polygons of an even number of sides inscribed in a conic.

We begin by recording a lemma, whose proof may be found in any reasonably exhaustive work on plane curves.

Lemma. If rn of the mn intersections of a curve of order m with a curve of order n lie on a curve of order r, then the remaining $(m-r)n$ *intersections lie on a curve of order* $m-r$.

Taking first the case when $m = n = 4$ and $r = 2$, the lemma reduces to the following:

If eight of the sixteen intersections of two quartic curves lie on a conic, then the other eight also lie on a conic.

This result remains valid if one or both of the quartics in question are reducible. For example, if both quartics break up into a pair of conics, one obtains a configuration of 16 points and six conics in the plane, each point lying on three of the conics and each conic containing eight of the points. Or again, both quartics may break up into four lines and we then have a configuration such as that illustrated in Figure 18.

Another interpretation of the same result yields an interesting property of octagons inscribed in a conic. Thus, suppose that 1, 2, . . . , 8 are the vertices of such an octagon (see Figure 19) and consider the two reducible quartic curves made up of the tetrads of lines 12, 34, 56, 78 and 23, 45, 67, 81. Eight of the intersections of these quartics are the points 1, 2, . . . , 8 with which we started. Since these lie on a conic, the other eight (denoted by 1′, 2′, . . . , 8′ in Figure 19) also lie on a conic. We therefore have the following:

Theorem of the inscribed octagon. If an octagon is inscribed in a conic, then the points of intersection of each side with the third side cyclically beyond it are the vertices of another octagon inscribed in a conic.

Figures 18 and 19, it may be noted, are projectively different figures; both, to be sure, illustrate the result about the 16 intersections of two tetrads of lines, but only Figure 19 illustrates the theorem of the inscribed octagon.

An interesting fact about the theorem of the inscribed octagon is that it is self-converse. For (with the notation of Figure 19) if we take the vertices of the 'outer' octagon in the order 8′3′6′1′4′7′2′5′, we see that the third side cyclically beyond 8′3′ is 1′4′, and these sides intersect in the vertex 1 of the original octagon; and so on.

Finally, to exhibit the connection between Pascal's theorem and the theorem of the inscribed octagon, we note that they are both special cases of the following general result (corresponding to the values $n = 3$ and $n = 4$ respectively).

If a polygon of $2n$ sides is inscribed in a given conic (the sides of the polygon being termed alternately 'odd' and 'even' according to some definite convention) then the $n(n-2)$ points where an odd side meets a non-adjacent even side lie on a curve of order $n - 2$.

Proof. We apply the lemma given at the beginning of this section to two (reducible) curves of order n, one consisting of the odd sides of the $2n$-*gon*, the other of the even sides. Of the n^2 intersections of these two curves, $2n$ lie on the given conic (namely, the vertices of the $2n$-*gon*) so, by the lemma (with $m = n$ and $r = 2$), the remaining $n(n - 2)$ intersections—which are each where an odd side meets a non-adjacent even side—lie on a curve of order $n - 2$, as required.

SECTION 3

CHAINS OF CIRCLES

A *chain* of circles, for the purposes of the present section, will be taken to mean a sequence $S_1, \ldots, S_n$ of circles in which each member touches both its neighbours; that is to say, S_1 touches S_2, S_2 touches $S_3, \ldots$, and S_{n-1} touches S_n. We shall sometimes refer to such a chain as an *open chain* and use the term *closed chain* for a sequence in which, additionally, the last circle S_n touches the first circle S_1.

The common theme of the theorems in this last section is that they are all concerned with closed chains of six circles. They also share a rather more intriguing common feature, namely, that their statements require rather more care than is at first apparent!

3.1. THE SEVEN CIRCLES THEOREM

Our first chain, which we call the seven circles theorem, is very easy to understand in principle (see Figure 20). As we have just hinted, however, it is not quite so easy to give a precise statement of the result. Let us begin with the following provisional assertion.

Let a, b', c, a', b, c' be a closed chain of six circles, all touching a base circle ω, and suppose that their points of contact with ω are six distinct points A, B', C, A', B, C' respectively. Then, subject to a certain extra condition to be discussed below, AA', BB' and CC' are concurrent.

We may add immediately that, if the six points $A, B', \ldots, C'$ are distributed round the circle ω cyclically in the order indicated, then the conclusion of the theorem is valid without further qualification. Figure 20 illustrates this case, as also does Figure 23. The points may, however, be distributed round ω in various other orders; Figures 21 and 22 illustrate two such possibilities, the conclusion of the theorem being valid in the former case but not in the latter.

Yet another case, in which the conclusion of the theorem is valid, is illustrated by the frontispiece of this book.

To gain some insight into the problem involved here, one should look at all the illustrations of this theorem, *temporarily disregarding the last circle c' of the chain.* It will then be noticed that, in each figure, there are two circles which touch a, b and ω and that the conclusion of the theorem is valid when c' is taken to be one of these, but not when it is taken to be the other. One is tempted to say that the theorem has a 50% chance of being true! In all the illustrations except Figure 22, the correct choice for c' has been made; in Figure 22, however, we have made the wrong choice. The reader should have no difficulty in spotting the other choice for c' in Figure 22 and it will be seen that, for this other choice, the lines AA', BB' and CC' are indeed concurrent. If one studies a number of different figures for the theorem, it soon becomes clear

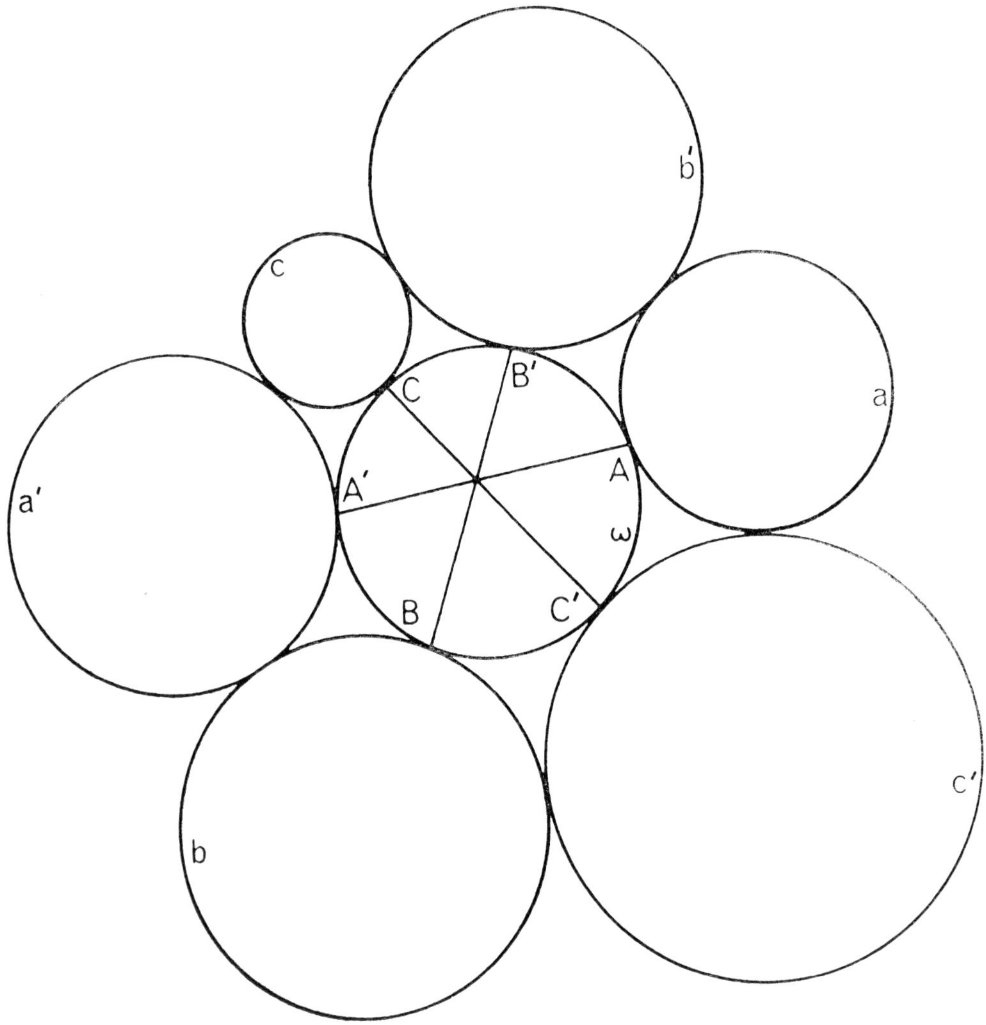

Figure 20 (see Section 3.1)

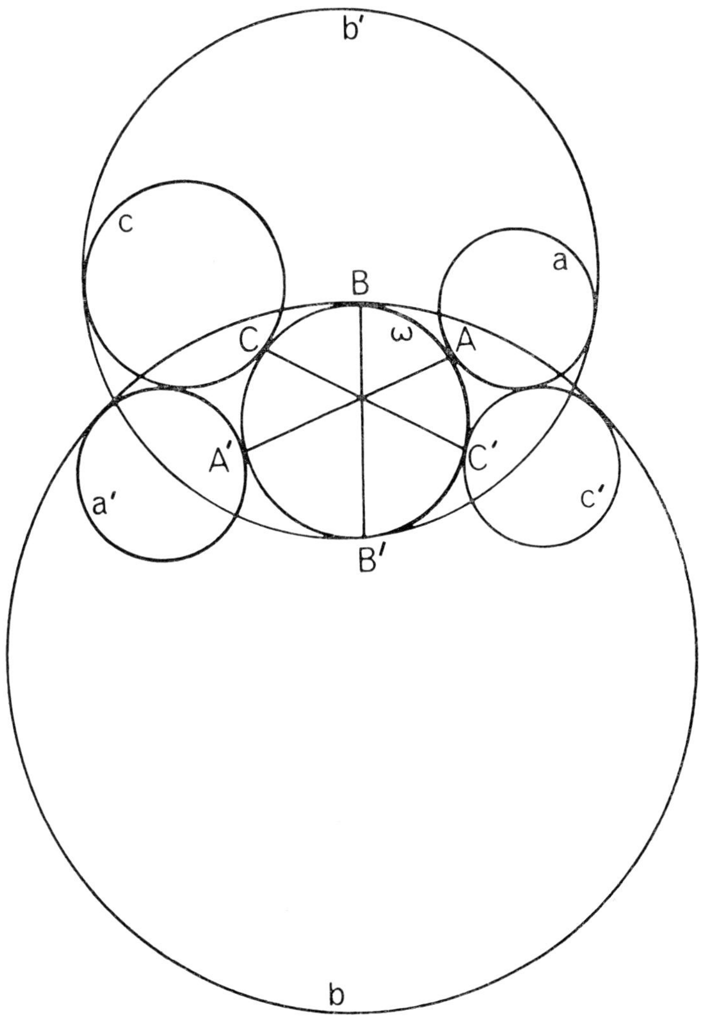

Figure 21 (see Section 3.1)

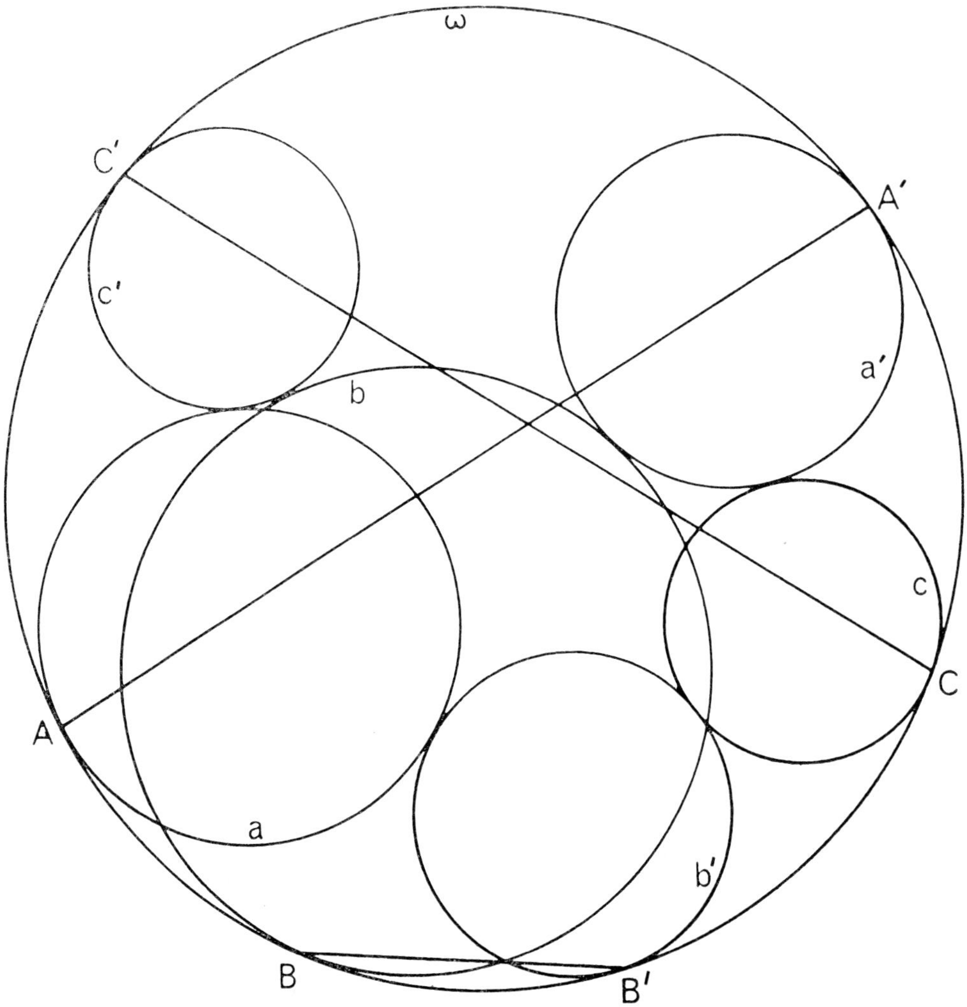

Figure 22 (see Section 3.1)

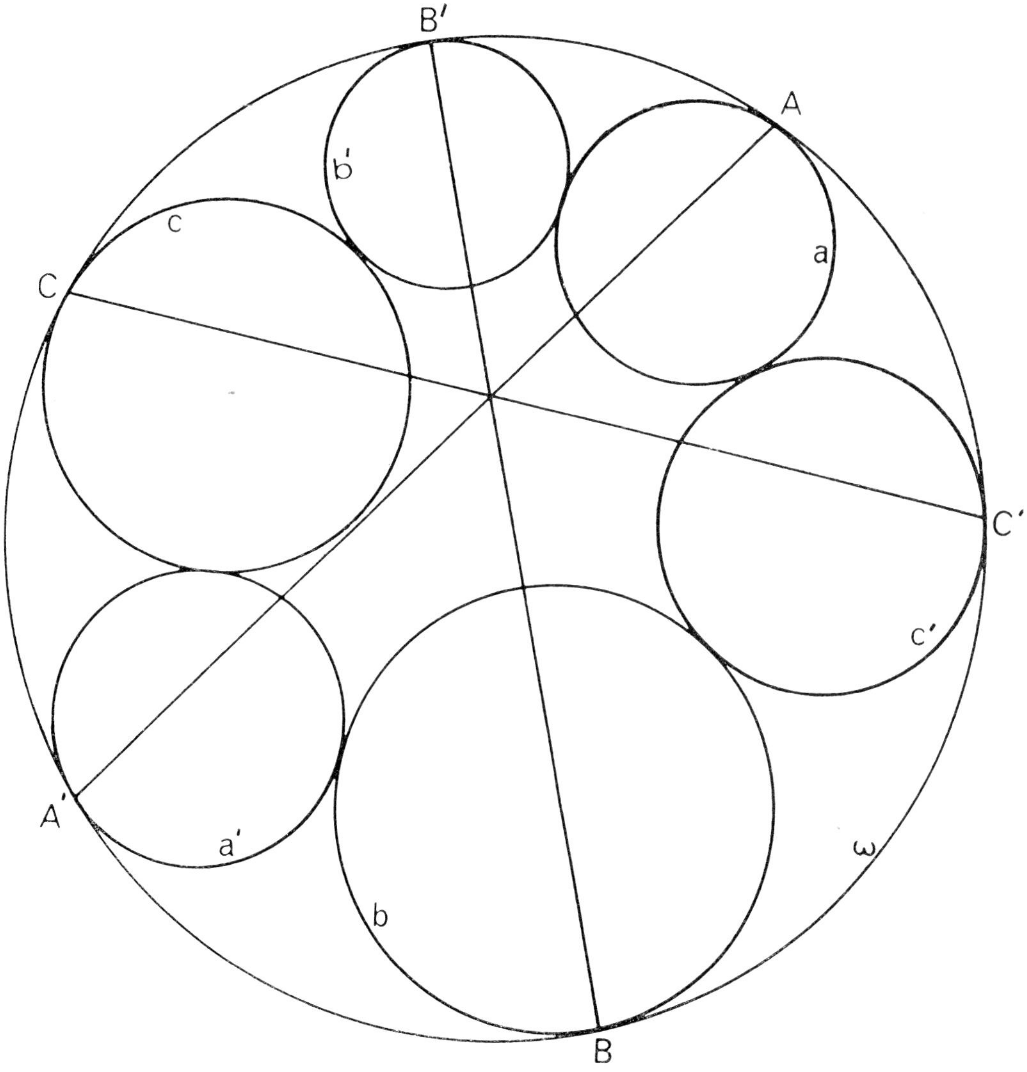

Figure 23 (see Section 3.1)

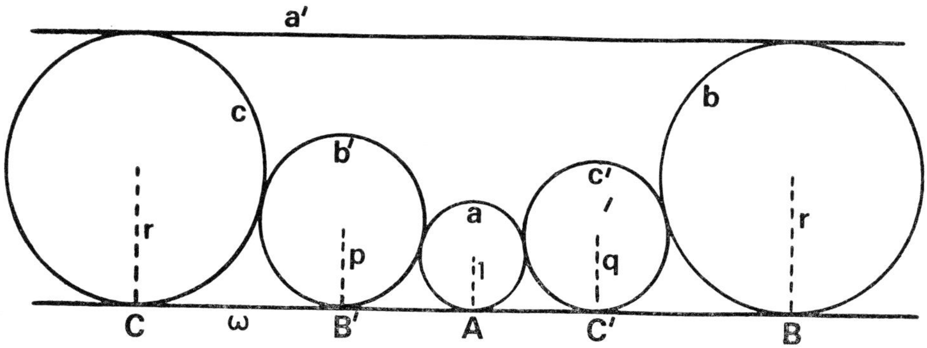

Figure 24 (see Section 3.1)

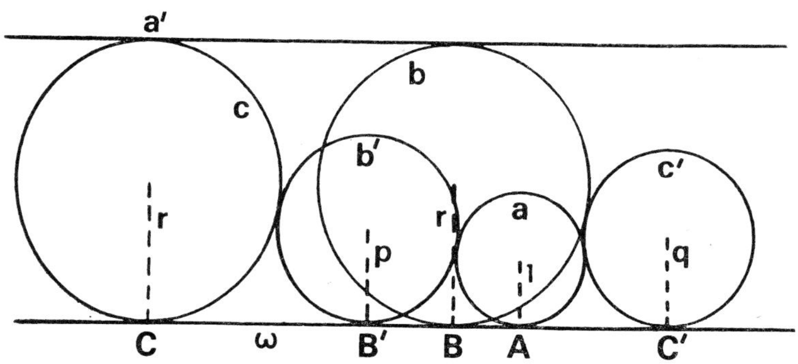

Figure 25 (see Section 3.1)

that its validity is dependent on some condition concerning the relative order of the six points A, B', . . . , C' round ω.

To give a precise formulation of this condition, we introduce the notion of a *positive arc* as follows: let P, Q, R be three points on a circle; there are, then, two arcs of the circle from P to R, only one of which goes through Q; we denote this arc by (PQR) and call it a positive arc if, in traversing it from P to R, we move round the circle in the positive (anti-clockwise) sense. With this understanding, we can now complete the statement of the seven circles theorem.

If, of the four arcs $(BC'A')$, $(C'AA')$, $(AB'A')$, $(B'CA')$, an even number are positive arcs, then the theorem as stated above is valid.

Proof.

We invert with respect to one of the six points of contact, A' say, and, as there is little fear of confusion, we use the same lettering for the inverse figure as for the original. In the inverse figure:

(i) a' and ω are parallel lines (being the inverses of circles which touch at A'),

(ii) b and c, which touch both a' and ω, are *equal* circles, and

(iii) b, c', a, b', c is an open chain of five touching circles, all touching ω, with points of contact B, C', A, B', C respectively.

Figure 24 illustrates one possibility for the inverse figure. Various other rather different-looking inverse figures are possible, but we may note that they all possess the following two features:

(a) All the circles in the inverse figure lie to the same side of ω; otherwise, there would be two touching circles in the figure which touched ω on opposite sides, and this would imply that they touched ω at the same point, contrary to our assumption (in the original statement of the theorem) that A, B', C, A', B, C' are distinct points. The corresponding feature of the figure before inversion is that the circles a, b', . . . , c' either all lie outside ω or all lie inside ω.

(b) All the contacts in the open chain b, c', a, b', c are external contacts (although this may not have been the case in the original figure). This follows from the observation that two circles with internal contact cannot touch a line ω at distinct points.

The reader may find it useful to refer to Figure 24 as an aid to the following proof, but it should be pointed out that the proof is valid quite generally, working equally well for all the different-looking inverse figures that may occur.

The centre of inversion has not been shown in Figure 24; its position is irrelevant to what follows. We do, however, need a name for this point, and we shall call it O. (It would be wrong, of course, to call it A', for A' has now been inverted into the point at infinity). Now the three lines AA', BB', CC' of the original figure invert respectively into the line OA, the circle OBB' and the circle OCC' and, since a pencil of straight lines inverts (except with respect to their point of concurrence) into a coaxal family of circles, our theorem will be established if we can show that OA is the radical axis of the circles OBB' and OCC'. Moreover, since O is already a common point of these two circles, it will be enough to prove that A is on their radical axis, i.e. that

$$\overline{AB'}.\ \overline{AB} = \overline{AC}.\ \overline{AC'}, \qquad (1)$$

these being the powers (squared tangential distances) of A with respect to the two circles. The bars in (1) are intended to signify that the segments concerned are *directed segments* on the line ω, and, in order to prove our theorem, we have to prove (1) as regards both magnitude and sign.

Now, still paying regard to sign, equation (1) may be written in the form

$$\overline{AB'}.\,(\overline{AC'} + \overline{C'B}) = (\overline{AB'} + \overline{B'C}).\,\overline{AC'}$$

so that, by removal of a common term from either side, the formula to be proved reduces to

$$\overline{AB'}.\,\overline{C'B} = \overline{B'C}.\,\overline{AC'}. \tag{2}$$

We show first that the left and right hand sides of (2) agree as to sign. The hypothesis made in the completion of the statement of the theorem was that an even number of the arcs $(BC'A')$, $(C'AA')$, $(AB'A')$, $(B'CA')$ are positive arcs (in the original figure); after inversion, with A' inverting to infinity, this means that an even number of the segments

$$\overline{BC'},\ \overline{C'A},\ \overline{AB'},\ \overline{B'C}$$

are positively directed segments on ω (where it is of no consequence which convention we make as to the positive direction on ω). It follows that an even number of the directed segments occurring in (2) are positive and hence that (2) is certainly correct so far as sign is concerned.

It remains then to show that the left and right hand sides of (2) are equal in magnitude. To see this, we let the radii of a, b' and c' (in the inverted figure) be 1, p and q respectively and we let the common radius of b and c be r. Now it is a fact of elementary geometry that, if two circles of radii r_1, r_2 touch externally, then the length of each common tangent is $2\sqrt{(r_1r_2)}$. Hence, since any two neighbouring circles of the open chain b, c', a, b', c touch externally, we have

$$BC' = 2\sqrt{(qr)},\ C'A = 2\sqrt{q},\ AB' = 2\sqrt{p},\ B'C = 2\sqrt{(pr)},$$

where we are now ignoring questions of sign. Using these values, we see that each side of equation (2) is equal in magnitude to the positive value of $4\sqrt{(pqr)}$ and our theorem is accordingly established.

It may be noted, incidentally, that equation (2) still holds, *so far as magnitude only is concerned*, even when (as in Figure 25) we invert a 'wrong' case of the seven circles theorem (Figure 22). The failure of the theorem in such a case is due to the fact that the left and right hand sides of (2) then have opposite signs.

It may also be noted that (in the original theorem) the point of concurrence of AA', BB' and CC' is the Brianchon point of the hexagon formed by the tangents to ω at A, B', . . . , C'. This follows at once from the well-known theorem that, if two chords AA' and BB' of a circle (or, more generally, of any conic) meet at X, then X lies on the line joining the pole of AB' to the pole of $A'B$.

We note finally that the statement that AA', BB' and CC' are concurrent may be alternatively expressed by saying that (A,A'), (B,B') and (C,C') are three pairs in involution on ω; and, using the statement in this alternative form, the seven circles theorem still has a non-trivial conclusion if ω is allowed to degenerate into a straight line. Other special cases of the theorem may be ob-

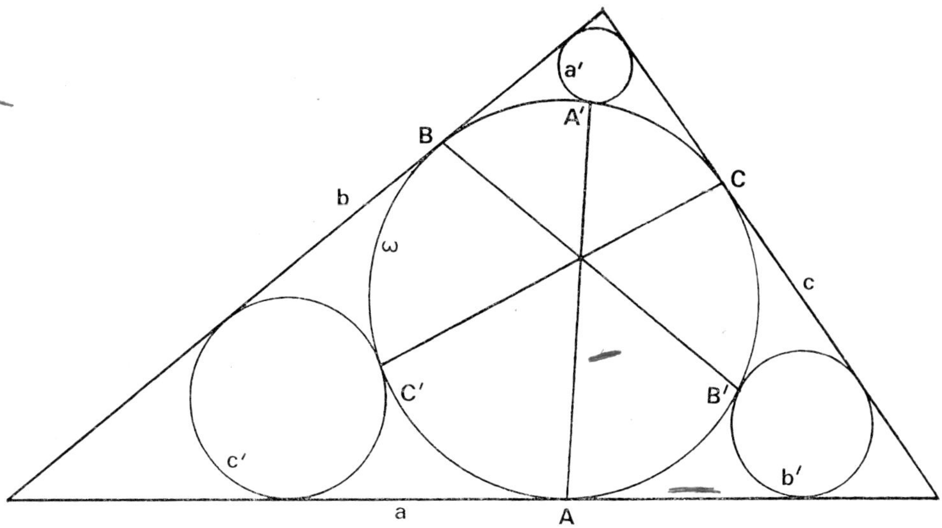

Figure 26 (see Section 3.1)

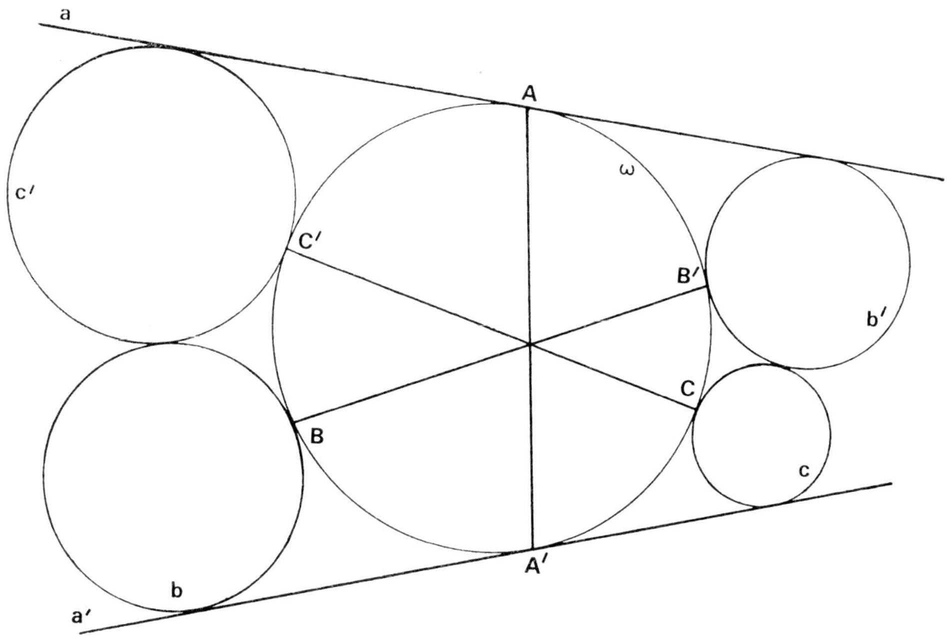

Figure 27 (see Section 3.1)

D

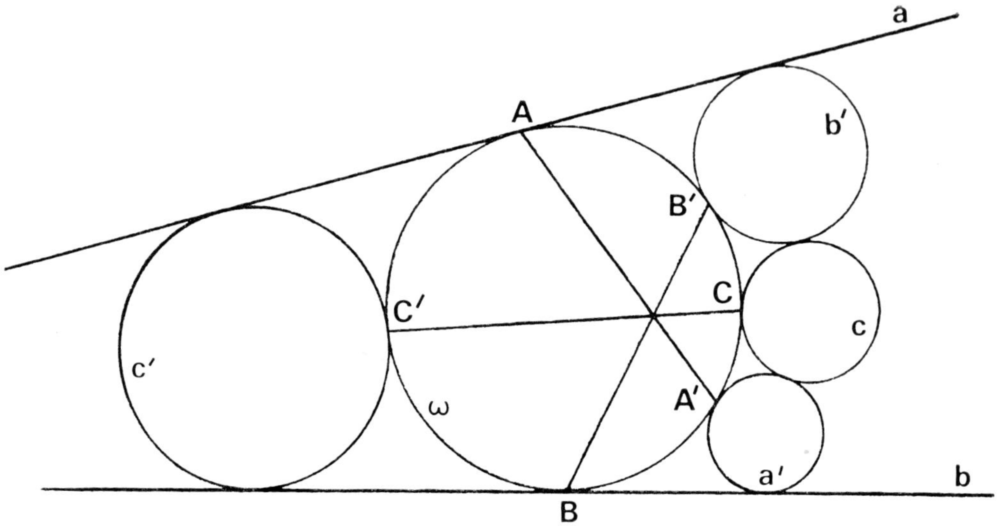

Figure 28 (see Section 3.1)

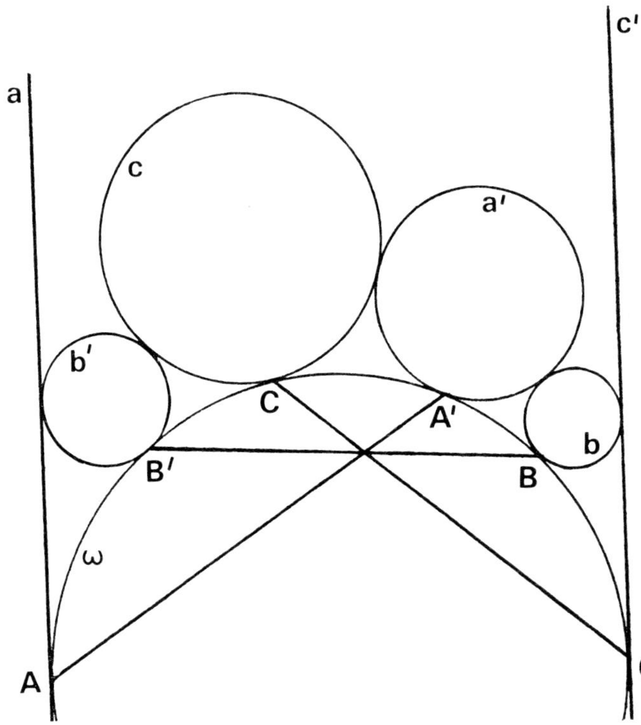

Figure 29 (see Section 3.1)

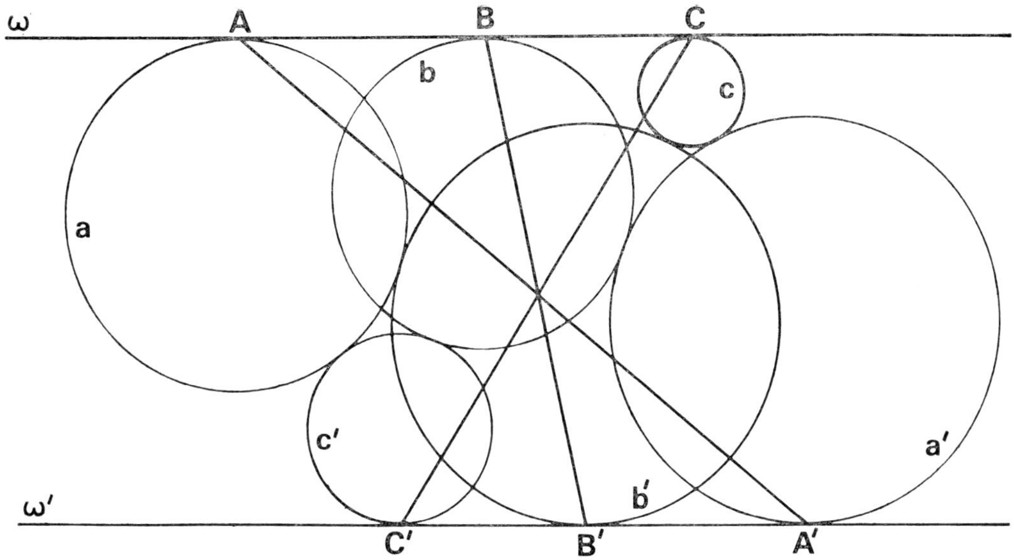

Figure 30 (see Section 3.2)

tained by allowing one or more of a, b', . . . , c' to degenerate into lines and we have included a selection of these in Figures 26-29. (Note that a is parallel to c' in Figure 29.)

3.2. A THEOREM ABOUT TWO PARALLEL STRAIGHT LINES AND SIX CIRCLES

The theorem to be discussed in this section will be seen to have certain affinities with the seven circles theorem of the last section. It would indeed be interesting to find a more general theorem of which both are special cases, but we have so far been unable to discover a simple theorem of this kind.

The idea of the present theorem is to replace the circle ω of the seven circles theorem by a pair of parallel straight lines ω and ω' and to consider a closed chain a, b', c, a', b, c' of circles which touch ω and ω' alternately (see Figure 30); it is then found that, subject to certain restrictions, the conclusion of the seven circles theorem carries over unchanged, i.e. that the lines AA', BB', CC' joining 'opposite' points of contact are concurrent. We must again be careful, however, to state the theorem correctly. Thus, referring to Figure 30, it may be seen that, for the given position of the circle a, it would be possible to take b' to be a rather larger circle crossing over the line ω and then there would be a number of different possibilities for c; namely it could touch b' either internally or externally and, in either case, it could touch ω either from above or from below.† To cut down this proliferation of possibilities, we focus attention on the case when

(a) all the contacts in the chain are external,

(b) the circles a, b, c touch ω from below, and

(c) the circles a', b', c' touch ω' from above.

Under these conditions, it turns out that the theorem is true; there are, in fact, rather more general conditions under which the theorem is also true, but we defer discussion of these until the end of this section. Let us, then, state the theorem to be proved.

If a, b', c, a', b, c' is a closed chain of externally touching circles which alternately touch two parallel straight lines ω and ω' (the former from below and the latter from above) and if their points of contact with the lines in question are A, B', C, A', B, C' respectively, then AA', BB' and CC' are concurrent.

Proof.

We let $2k$ denote the distance between ω and ω'. Our proof will depend on the repeated application of the following simple trigonometrical lemma.

Lemma. Let a be a circle with centre X touching ω from below and let b' be a circle with centre Y' touching ω' from above; and suppose that a and b' touch externally. If the horizontal projection of XY' is then denoted by $2kp$, we assert that the vertical projection of XY' is $k(1 - p^2)$. This formula, moreover, gives the vertical projection of XY' both in magnitude and sign, being a positive quantity if Y' is at a lower horizontal level than X and negative in the contrary case.

Proof of the lemma. We let x and y' denote the radii of a and b' respectively

†We trust that the reader will understand the phrases 'from above' and 'from below' as convenient abbreviations for descriptions which would otherwise be unjustifiably pedantic. Various other informal phrases of this kind will also be used in this section.

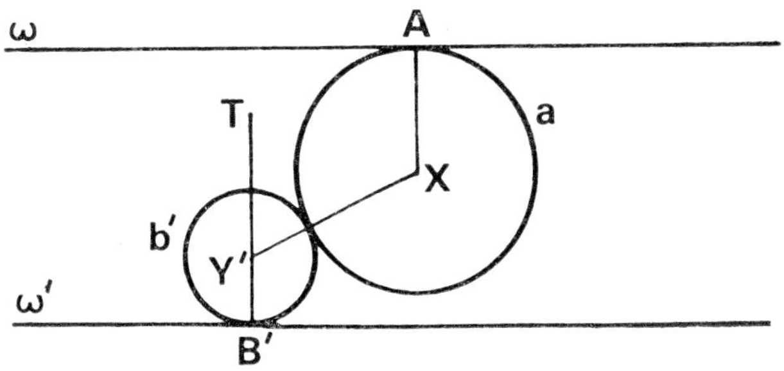

Figure 31a (see Section 3.2)

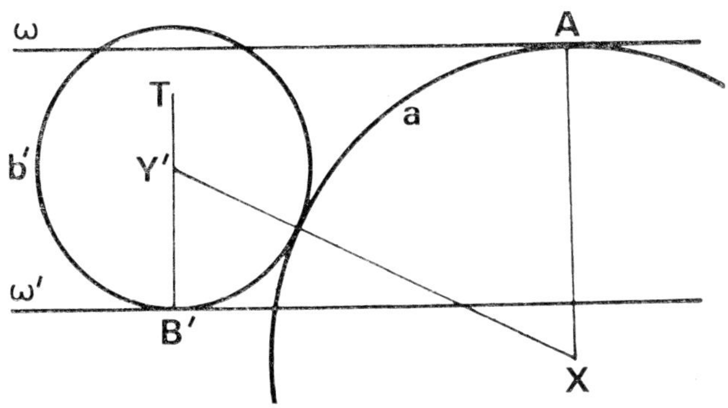

Figure 31b (see Section 3.2)

and we let A and B' be the points of contact with ω and ω'. We also let θ denote the angle $TY'X$, where T is any point on $B'Y'$ produced. Figures 31a and 31b illustrate the two possible cases of the lemma; in the first of these, θ is acute, Y' is at a lower horizontal level than X and the vertical projection of XY' is positive (according to the convention stated above); in the second figure, θ is obtuse, Y' is at a higher horizontal level than X and the vertical projection of XY' is negative.

Consider now, in either figure, the polygonal arc $AXY'B'$. Since the vertical projection of this arc must have length equal to $2k$ (the distance between ω and ω') we have

$$x + (x + y') \cos \theta + y' = 2k,$$

that is

$$x + y' = \frac{2k}{1 + \cos \theta}. \tag{1}$$

Now the horizontal and vertical projections h and v of XY' are

$$h = (x + y') \sin \theta \quad \text{and} \quad v = (x + y') \cos \theta,$$

the latter being positive if θ is acute and negative if θ is obtuse. Using (1), we therefore have

$$h = \frac{2k \sin \theta}{1 + \cos \theta}, \quad v = \frac{2k \cos \theta}{1 + \cos \theta}.$$

If, then, we put $p = \tan \frac{1}{2}\theta$ and use the well-known trigonometric identities

$$\sin \theta = \frac{2p}{1 + p^2}, \quad \cos \theta = \frac{1 - p^2}{1 + p^2},$$

the formulae for h and v reduce to

$$h = 2kp, \ v = k(1 - p^2)$$

and the lemma is established.

Returning now to the proof of the main theorem, we let X, Y', Z, X', Y, Z' denote the centres of the circles a, b', c, a', b, c' respectively. We also denote the horizontal projections of XY', $Y'Z$, ZX', $X'Y$, YZ', $Z'X$ by $2kp$, $2kq$, $2kr$, $2ks$, $2kt$, $2ku$ respectively. It was unnecessary, in the lemma, to have a sign convention for horizontal distances; however, we now find it convenient to introduce such a convention; namely, we suppose that p is positive or negative according as Y' is to the right or to the left of X; and we make similar conventions for $q, r, \ldots, u$. Since $XY'ZX'YZ'X$ is a closed polygon, the properly signed horizontal projections of $XY', \ldots, Z'X$ will add up to zero, i.e.

$$p + q + r + s + t + u = 0. \tag{2}$$

The properly signed vertical projections will also add up to zero and, by the lemma, this gives

$$\begin{aligned} k(1 - p^2) - k(1 - q^2) + k(1 - r^2) \\ - k(1 - s^2) + k(1 - t^2) - k(1 - u^2) = 0 \end{aligned} \tag{3}$$

where, it will be observed, the terms are alternately positively and negatively signed; the reason for this should be apparent to the reader. Now equation (3)

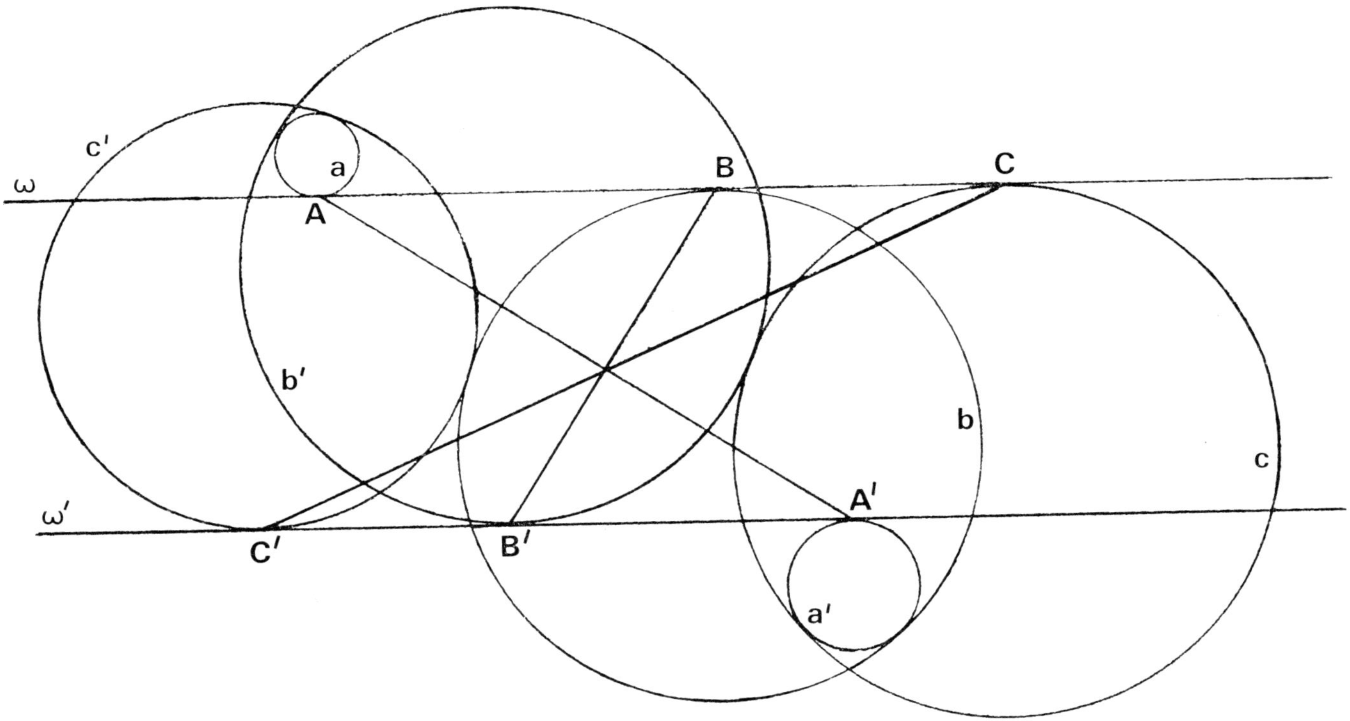

Figure 32 (see Section 3.2)

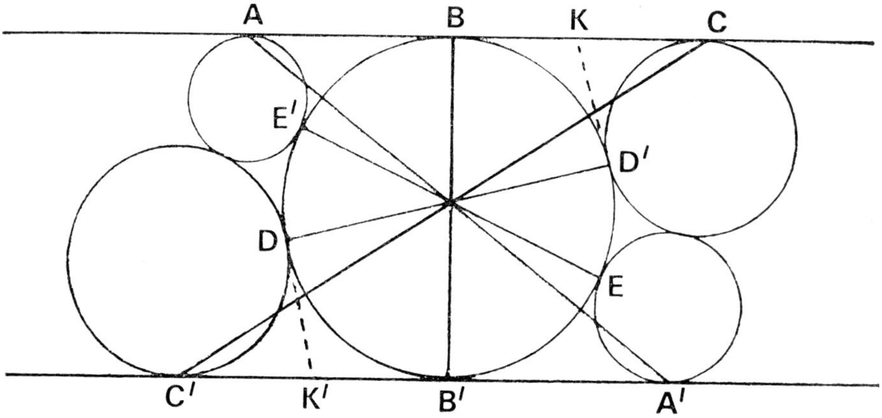

Figure 33 (see Section 3.2)

may be conveniently written in the form

$$r^2 - q^2 - s^2 = u^2 - p^2 - t^2; \tag{4}$$

also, by equation (2), we have

$$(r + q + s)^2 = (u + p + t)^2. \tag{5}$$

If we add (4) and (5) and divide through by a factor 2, we find that

$$(r + s)(q + r) = (u + p)(t + u),$$

that is

$$\frac{r+s}{t+u} = \frac{u+p}{q+r}. \tag{6}$$

Now $r + s$ is the sum of the horizontal projections of ZX' and $X'Y$ and is therefore the horizontal projection of ZY. Moreover, since C and B lie vertically above Z and Y respectively, it follows that $r + s$ is the properly signed distance CB. Similarly, we find that $t + u$, $u + p$ and $q + r$ are the properly signed distances BA, $C'B'$ and $B'A'$. Equation (6) is therefore equivalent to

$$\frac{CB}{BA} = \frac{C'B'}{B'A'}$$

and now, since A,B,C and A',B',C' lie on parallel lines ω and ω', it follows by similar triangles that AA', BB' and CC' are concurrent; our theorem is therefore proved.

We now remark that, with the exception of the opening lemma, the above proof consists only of straightforward algebraic manipulation; the proof will therefore remain valid in all cases for which the conclusion of the lemma is valid (whether or not the conditions we imposed as to external contact, etc., are satisfied). Now it so happens that the lemma remains valid under rather less strict conditions than those that were stated earlier; more precisely, if $2kp$ is the horizontal projection of XY', then its vertical projection is given (with correct sign) by $k(1 - p^2)$ in each of the following three cases:

(i) if one circle touches ω from below and the other touches ω' from above, with external contact between the circles;

(ii) if one circle touches ω from above and the other touches ω' from above, with internal contact between the circles;

(iii) if one circle touches ω from below and the other touches ω' from below, with internal contact between the circles.

We may leave it as an exercise for the reader to verify this result for cases (ii) and (iii); our own proof was, of course, restricted to case (i) only. Since, as we have just said, the proof of our theorem subsequent to the lemma was only straightforward algebra, we can now assert that:

If a, b', c, a', b, c' is a closed chain of circles which alternately touch two parallel straight lines ω and ω' at A, B', C, A', B, C' respectively, and if every pair of neighbouring circles in the chain satisfy one or other of the three conditions (i), (ii) and (iii) given above, then AA', BB' and CC' are concurrent.

Figure 32 illustrates a case of this slightly more general result; in this figure, the pairs of circles (b,c') and (b',c) satisfy condition (i), the pairs (a,b') and

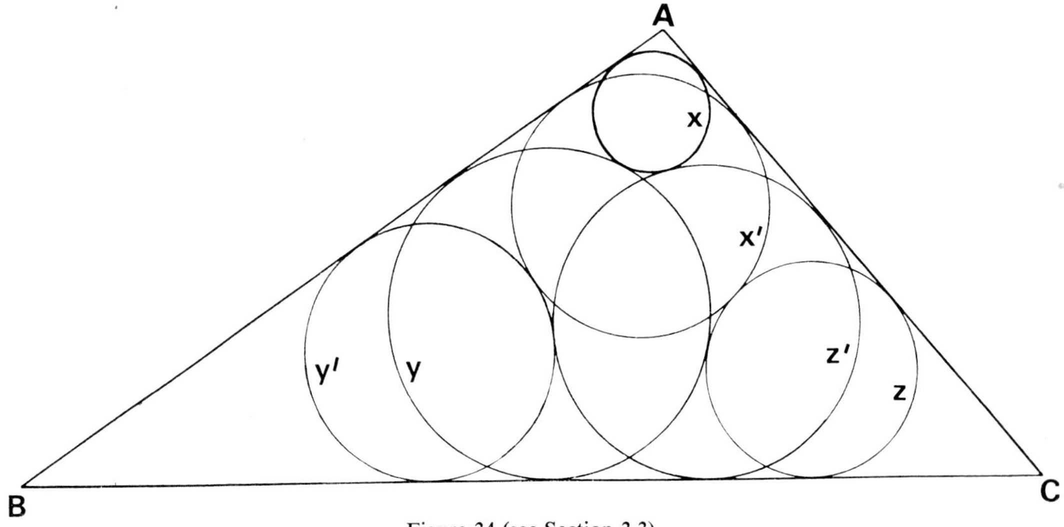

Figure 34 (see Section 3.3)

(a,c') satisfy condition (ii), while the pairs (a',b) and (a',c) satisfy condition (iii).

Finally, in this section, we draw attention to the rather charming Figure 33 (in which the horizontal lines are parallel). Firstly, as a special case of the theorem of the present section, the lines AA', BB' and CC' are concurrent at X, say; and secondly, as a special case of the seven circles theorem, the lines BB', DD' and EE' are concurrent at Y, say. It can then be shown further that *X and Y are the same point*. We leave the proof of this as an exercise for the reader. (Hint. Let K and K' be the midpoints of BC and $B'C'$ and prove that the point where KK' meets BB' coincides with both X and Y.)

3.3. A THEOREM ABOUT A TRIANGLE AND SIX CIRCLES

We turn now to a chain theorem associated with a triangle ABC (Figure 34). The basic idea is to start with an arbitrary circle x which touches the sides CA and AB of the triangle and then to build up what we may call a *triangular chain* as follows:

y is a circle touching AB, BC and x,
z is a circle touching BC, CA and y,
x' is a circle touching CA, AB and z,
y' is a circle touching AB, BC and x',
z' is a circle touching BC, CA and y',
x'' is a circle touching CA, AB and z'.

These specifications do not, of course, determine the circles $y, z, \ldots, x''$ uniquely; various choices are available at each stage; but, as we shall show in this section, *if the choice is properly made at each stage, then the last circle x'' coincides with the first circle x (and we then have a closed triangular chain of six circles).*

As may be observed, this theorem is closely related to the classical problem of Malfatti (which is to choose x in such a way that x' coincides with it); but, whereas Malfatti's problem has only a finite number of solutions, the present chain of circles closes up for arbitrary choice of x.

One way of cutting down the number of available choices at each stage is to insist that the circles of the chain shall lie *inside the triangle ABC.* This, as is easily seen, gives *at most one choice*† for each successive circle in the chain and, in addition, corresponds to the most natural way of drawing the figure for the theorem. We therefore begin by proving the theorem in this case, deferring consideration of the other possible cases until later on. We state:

If x, y, z, x', y', z', x'' is a triangular chain of circles inscribed in the angles of a triangle ABC, and if all the circles of the chain lie inside the triangle, then x'' coincides with x.

Proof.

We begin by noting two consequences of our assumption that the circles lie inside the triangle.

†There is no guarantee, however, that there is any choice at all. Thus, if ABC is obtuse-angled at B and if the first circle x is a sufficiently small circle inscribed in the angle A, then there is no choice for *y inside the triangle*.

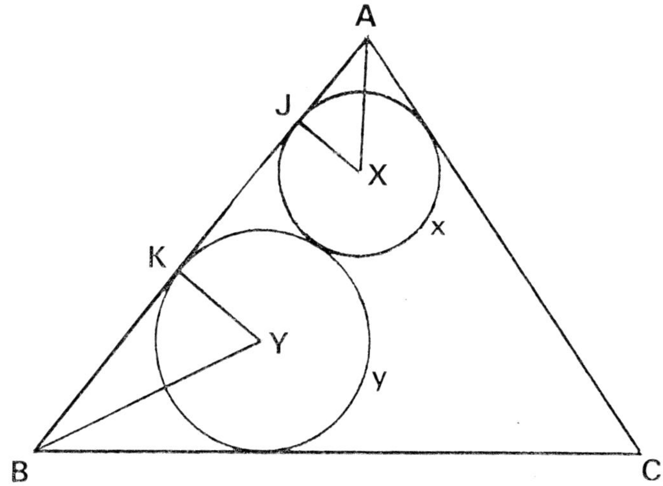

Figure 35 (see Section 3.3)

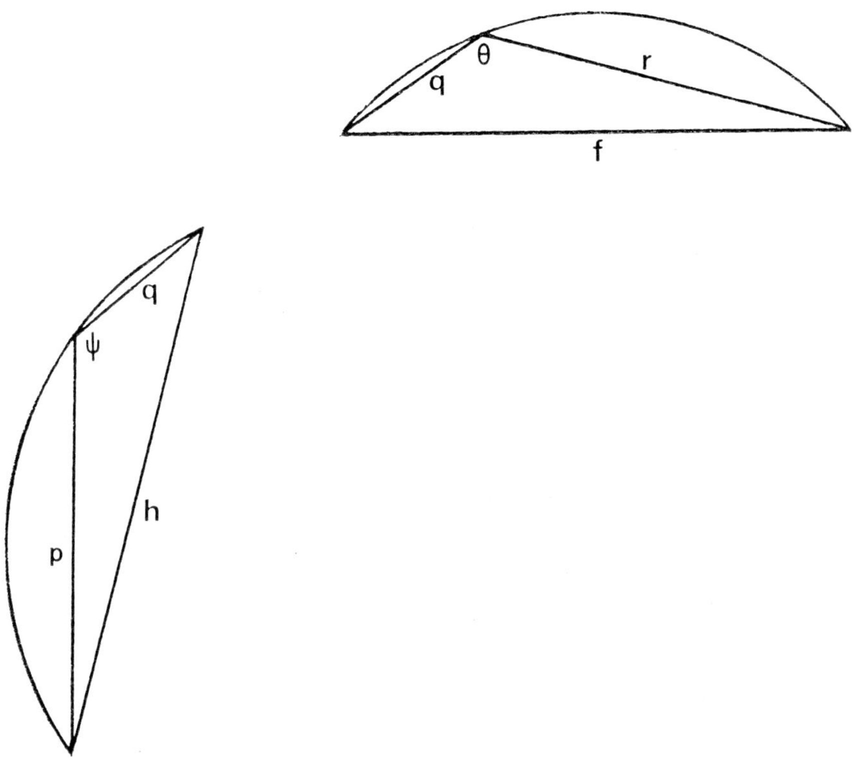

Figure 36 (see Section 3.3)

(i) The centres of the circles of the chain must lie on the appropriate *internal* angle-bisectors of the triangle.

(ii) All the contacts between successive circles of the chain must be *external* contacts.

As to notation, we let a, b and c, as usual, denote the lengths of the sides BC, CA and AB of the triangle and we let l, m, n, l', m', n', l'' denote the lengths of the tangents to x, y, z, x', y', z', x'' from A, B, C, A, B, C, A respectively. The idea of the following proof is to obtain the relations connecting consecutive pairs of these lengths $l, m, \ldots, l''$ and thence, by eliminating m, n, l', m' and n', to deduce that $l = l''$, which will evidently prove the theorem.

First, then, we obtain the relation between l and m. Figure 35 shows the two circles x and y in the required position, their centres being X and Y and their points of contact with AB being J and K. By definition, we have $AJ = l$ and $BK = m$. Moreover, since X is on the internal bisector of the angle A, we have $JX = l \tan \frac{1}{2}A$, this being the radius of x. Similarly, the radius of y is $KY = m \tan \frac{1}{2}B$. Now, as we have had occasion to remark earlier (in Section 3.1), if two circles of radii r_1, r_2 touch externally, the length of each common tangent is $2\sqrt{(r_1 r_2)}$; thus, in the present case, we have

$$JK = 2\sqrt{(lm \tan \tfrac{1}{2}A . \tan \tfrac{1}{2}B)}.$$

Hence, since $AB = AJ + BK + JK$, we obtain

$$c = l + m + 2\sqrt{(lm \tan \tfrac{1}{2}A . \tan \tfrac{1}{2}B)}. \qquad (1)$$

Now, by the familiar half-angle formulae for a triangle, we have

$$\tan \tfrac{1}{2}A . \tan \tfrac{1}{2}B = \sqrt{\frac{(s-b)(s-c)}{s(s-a)}} \sqrt{\frac{(s-c)(s-a)}{s(s-b)}} = \frac{s-c}{s}$$

where $s = \frac{1}{2}(a + b + c)$. Equation (1) therefore reduces to

$$c = l + m + 2\sqrt{(lm)}\sqrt{(1 - c/s)} \qquad (2)$$

and this is the required relation connecting l and m. The corresponding relations connecting m and $n, \ldots, n'$ and l'' can now be written down by analogy. Thus

$$a = m + n + 2\sqrt{(mn)}\sqrt{(1 - a/s)},$$

and so on, there being six relations in all.

To eliminate m, n, l', m', n' between these six relations, we play the following amusing trick. We write p, q, r, p', q', r', p'' for the (positive) square roots of l, m, n, l', m', n', l'' and we write f, g, h and t for the square roots of a, b, c and s. We also define obtuse angles θ, ϕ, ψ by means of the relations

$$\begin{aligned}
\cos \theta &= -\sqrt{(1 - a/s)} = -\sqrt{(1 - f^2/t^2)}, \quad \sin \theta = f/t,\\
\cos \phi &= -\sqrt{(1 - b/s)} = -\sqrt{(1 - g^2/t^2)}, \quad \sin \phi = g/t,\\
\cos \psi &= -\sqrt{(1 - c/s)} = -\sqrt{(1 - h^2/t^2)}, \quad \sin \psi = h/t, \qquad (3)
\end{aligned}$$

this being evidently possible since the quantities under the square root signs are positive and between 0 and 1. Equation (2) thereby reduces to the form

$$h^2 = p^2 + q^2 - 2pq \cos \psi \qquad (4)$$

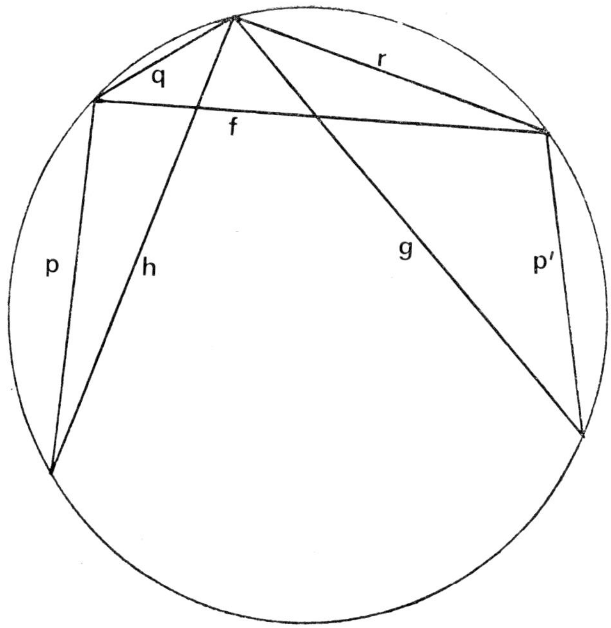

Figure 37 (see Section 3.3)

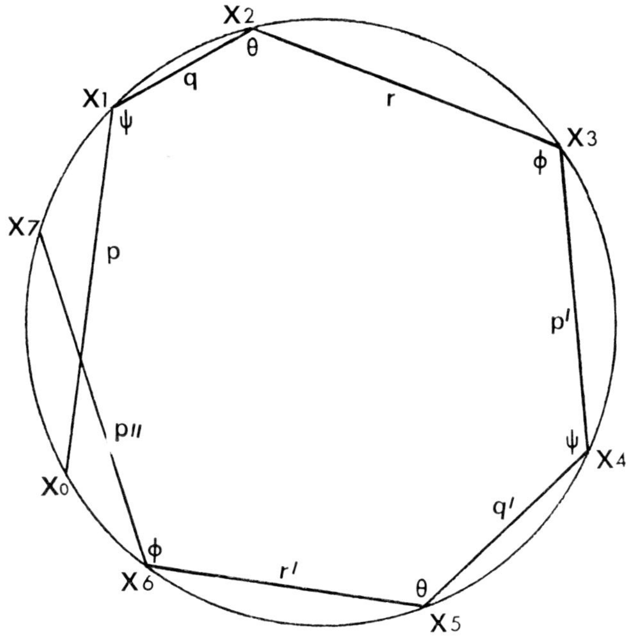

Figure 38 (see Section 3.3)

and the five other equations reduce to similar forms, with h, p, q, ψ replaced by the successive columns of the following table:

h	f	g	h	f	g
p	q	r	p'	q'	r'
q	r	p'	q'	r'	p''
ψ	θ	ϕ	ψ	θ	ϕ

Since our original aim was to prove that $l = l''$, the objective now is to eliminate q, r, p', q', r' between (4) and the five companion equations and to obtain $p = p''$.

Now equation (4), of course, expresses the familiar cosine rule and tells us that a triangle can be constructed with sides p, q and h having an angle ψ opposite to h; similarly, the five companion equations give us the specifications for five other triangles. We thus have six triangles in all, the first two of which are shown in Figure 36. Now the remarkable fact about these six triangles is that they all have the same circumradius; to see this, it is only necessary to recall that the diameter of the circumcircle of a triangle is equal to the length of any side divided by the sine of the opposite angle; so, for the first of our six triangles, the diameter in question is $h/\sin \psi$ which, by (3), is equal to t; and the same value t is obtained for the five other triangles. It follows that, if we fit the two triangles of Figure 36 together so that their sides of length q coincide, then their circumcircles also coincide. Figure 37 shows the two triangles fitted together in this way, with the third triangle put in as well. Proceeding thus, we eventually fit all six triangles together (into the same circumcircle) obtaining the situation of Figure 38, where certain lines have been left out in the interests of clarity.

To sum up the situation now reached, we have (in Figure 38) an open heptagon $X_0X_1 \ldots X_7$ inscribed in a circle, with angles ψ at X_1 and X_4, angles θ at X_2 and X_5, and angles ϕ at X_3 and X_6; and, of course, the aim is still to prove that $p = p''$, i.e. that X_0X_1 and X_6X_7 are of equal length. To prove this, we notice that the triangles $X_0X_2X_4$ and $X_3X_5X_7$ are congruent (because $X_0X_2 = X_3X_5 = h$, $X_2X_4 = X_5X_7 = g$, and the angles opposite these sides, being subtended in the same circle by chords of equal length, are equal). It follows that the angles $X_0X_2X_4$ and $X_3X_5X_7$ are equal. If we subtract these angles from the equal angles $X_1X_2X_3$ and $X_4X_5X_6$ respectively (which are both equal to θ) we are left with

$$\angle X_0X_2X_1 + \angle X_3X_2X_4 = \angle X_3X_5X_4 + \angle X_6X_5X_7. \quad (5)$$

The angles $X_3X_2X_4$ and $X_3X_5X_4$, however, are equal (in the same segment); so we deduce from (5) that

$$\angle X_0X_2X_1 = \angle X_6X_5X_7.$$

This shows that X_0X_1 and X_6X_7 subtend equal angles at the circumference of the circle and implies that these chords are equal in length. Hence $p = p''$ and our theorem is proved.

Returning now to the problem of describing the other cases in which a triangular chain closes up, let us relax the condition that the circles lie inside the triangle and, instead, require only that their centres lie on the internal bisectors of the respective angles of the triangle. It is then the case that, for a given choice of the first circle x, there are three possible choices for y (two of which can be non-real for certain positions of x).

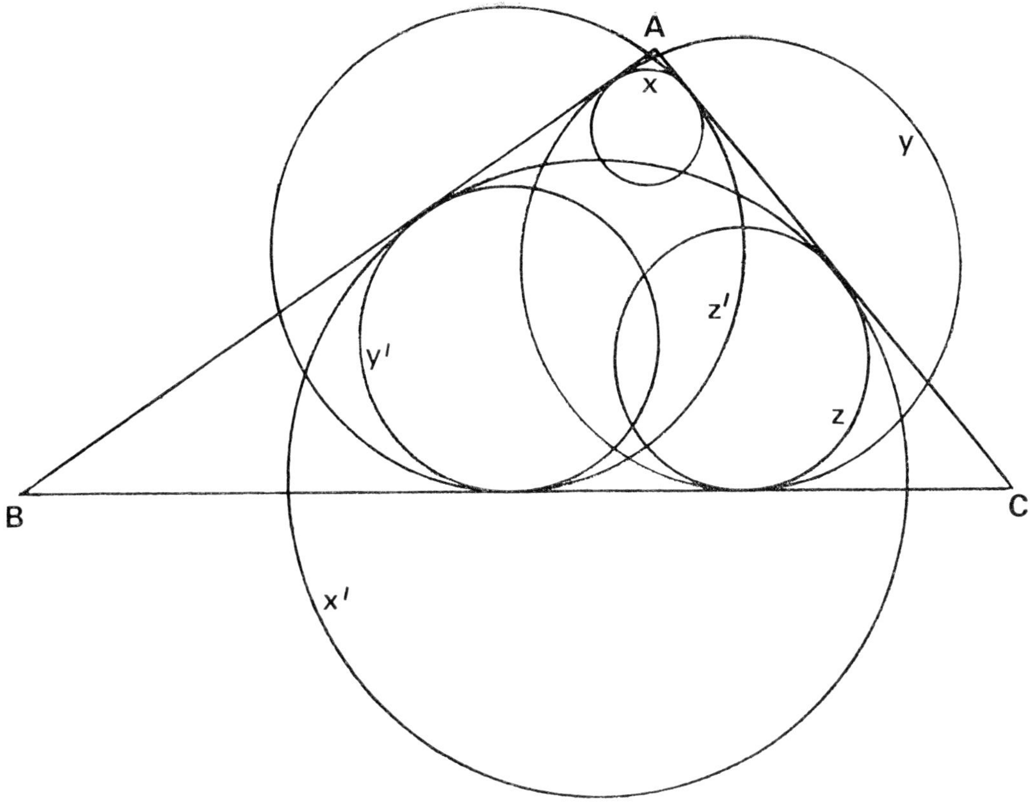

Figure 39 (see Section 3.3)

E

(a) One of these three, which we term the *special choice* for y (and which is always real) is rather different from the other two in that *it touches AB at the same point as* x (and consequently touches x internally). If we take this special choice for y, we can then also take the special choice for z (which touches BC at the same point as y), and so on. Such a chain, built up with special choices at every stage, necessarily closes up (see Figure 39). The proof of this is extremely easy for, letting $l, m, \ldots, l''$ have the meanings ascribed to them earlier in this section, we see at once that

$$l + m = c, \qquad m + n = a, \qquad n + l' = b,$$
$$l' + m' = c, \qquad m' + n' = a, \qquad n' + l'' = b.$$

Hence $l = l''$ and the result follows immediately.

(b) The other two choices for y (excluding now the special choice) turn out to be real if and only if x does not lie entirely outside the triangle; obviously they both touch x externally. It is easy moreover to show that, with the notation of the earlier part of this section, the quantities l and m still satisfy the equation (2)—except now that both values of the square root must be taken into account (corresponding to the fact that there are two possibilities for y). It is more appropriate, therefore, to write (2) in the rationalized form

$$s(c - l - m)^2 = 4(s - c)\, lm; \tag{6}$$

this is a quadratic for m (when l is given) and has real roots lying between 0 and s if and only if l lies between 0 and s. This confirms the statement made above concerning reality.

If we now construct a triangular chain, taking at each stage one of the two non-special choices for the next circle, then, after six stages, we obtain an apparent total of 64 possibilities for x'' corresponding to any given position of the first circle x; and our problem is to decide how many of these coincide with x. Algebraically, we have equation (6) and five similar relations corresponding to the later stages of the chain, and elimination of m, n, l', m', n' between these relations will lead to the determination of l'' as a 64-valued function of l; and the problem is to find how many of these 64 values of l'' are equal to l.

This problem is, in fact, susceptible of solution by the method used earlier for the case when all the circles are inside the triangle, but the following alternative method may also be of interest. Namely, we replace the parameters l, m, n, l', m', n', l'' by new parameters $\lambda, \mu, \nu, \lambda', \mu', \nu', \lambda''$ given by

$$l = s \sin^2 \lambda, \quad m = s \sin^2 \mu, \ldots, l'' = s \sin^2 \lambda'', \tag{7}$$

noticing that these new parameters are determined only to within sign (mod π). Omitting the straightforward calculation, it is found that, in terms of these new parameters, equation (6) reduces to $\cos 2\mu = \cos (2\lambda \pm 2\psi)$, where ψ is the angle introduced earlier (equations (3)). Consequently, we have

$$\mu = \pm \lambda \pm \psi \pmod{\pi}$$

and similarly, from the five other relations,

$$\nu = \pm \mu \pm \theta \pmod{\pi}, \ldots, \lambda'' = \pm \nu' \pm \phi \pmod{\pi}.$$

Hence

$$\lambda'' = \pm \lambda + (\pm \psi \pm \theta \pm \phi \pm \psi \pm \theta \pm \phi) \pmod{\pi} \tag{8}$$

where the successive sign alternatives within the brackets on the right hand side correspond to the successive choices that can be made as we construct the

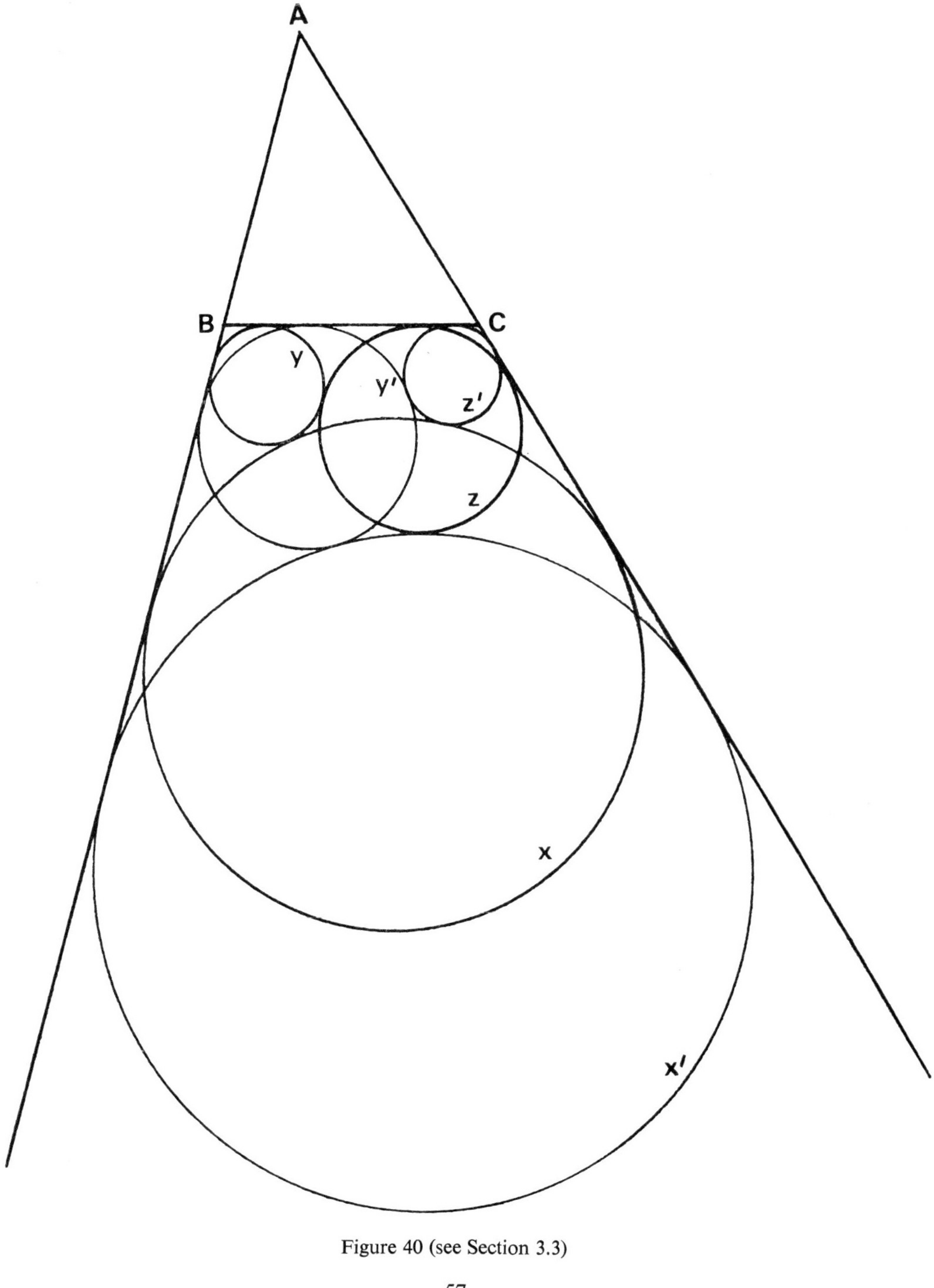

Figure 40 (see Section 3.3)

chain. The aim of course is to choose the sign alternatives in such a way as to get $\lambda'' = \pm\lambda$ (for this, by (7), will imply $l'' = l$). Now we see from (8) that, in order to achieve this, *we may make the first three sign choices quite arbitrarily provided that, thereafter, we make 'correct' choices so that the terms cancel off in pairs.* We conclude therefore that, of the 64 possible chains of non-special choices beginning with x, eight will close up. Of course, if there happens to be a chain of circles *lying inside the triangle* which begins with x, then it must be one of these eight.

We sum up the findings of our discussion in the following theorem.

Let I be the incentre of the triangle ABC and suppose we construct a triangular chain $x, y, \ldots$ of circles with centres lying on the appropriate angle bisectors AI, BI, CI. Then

(a) if a special choice is made at each stage, the chain closes up; and

(b) if a non-special choice is made at each stage, then each of x, y, z, x' may be chosen at will and there are then 'correct' choices for y' and z' so that the chain closes up.

As a final extension, we point out that the above result carries over word for word if I is taken to be the centre of an escribed circle of ABC instead of the incentre. Figure 40 illustrates a case of this kind.

3.4. THE NINE CIRCLES THEOREM

We come now to our final chain. This is a generalisation of the chain of the previous section obtained by replacing the sides of the triangle by a set of three circles. More precisely, let a, b and c be three circles in the plane and let x be any circle touching b and c; we then build up a chain as follows:

y is a circle touching c, a and x,
z is a circle touching a, b and y,
x' is a circle touching b, c and z,
y' is a circle touching c, a and x',
z' is a circle touching a, b and y',
x'' is a circle touching b, c and z'.

As in the triangle theorem, there are a number of choices for each successive circle in the chain but, as we shall show in this section, *if the choice at each stage is appropriately made (in a manner to be explained below) then the last circle x'' coincides with the first circle x.*

Figures 41 and 42 illustrate two examples of this chain and it may be noticed that the system of nine circles $a, b, c, x, y, z, x', y', z'$ is a symmetrical system, each circle being touched by four others. This symmetry may be put in evidence by considering the following array:

$$\begin{matrix} a & b & c \\ x & y' & z \\ x' & y & z' \end{matrix}$$

Thus, it will be observed that two of the nine circles touch if (and generally only if) their names do not appear in the same row or column of this array.

We now proceed to describe how the 'appropriate choices' mentioned above

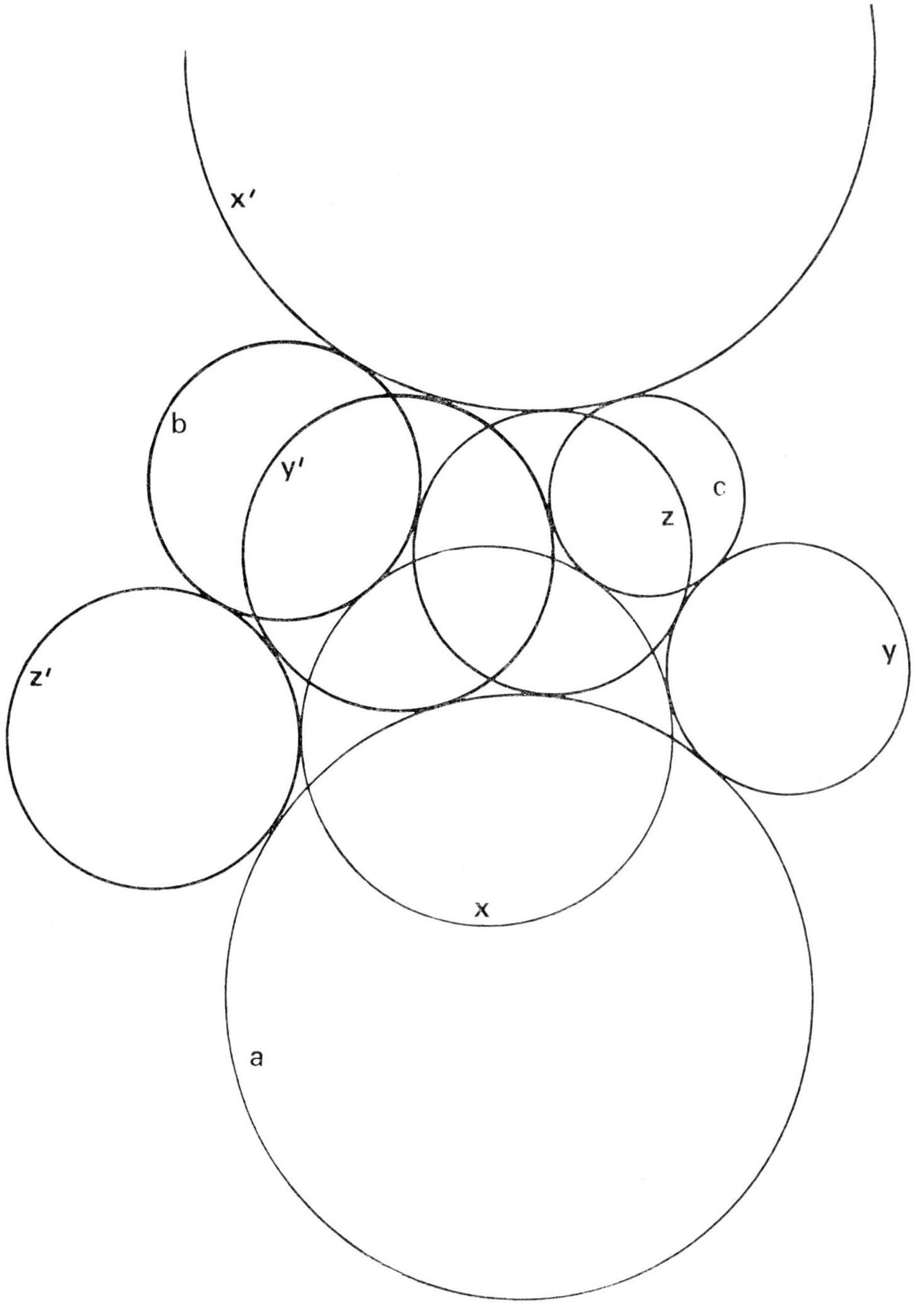

Figure 41 (see Section 3.4)

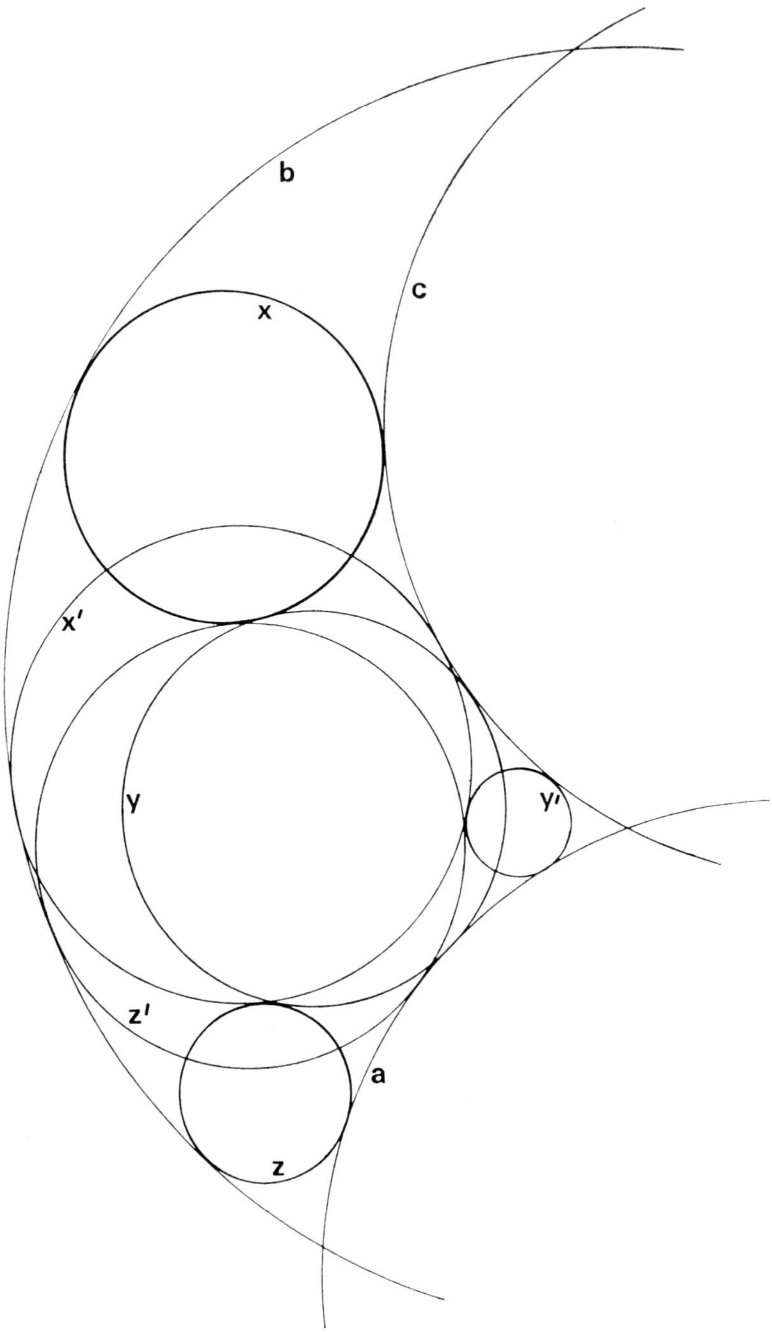

Figure 42 (see Section 3.4)

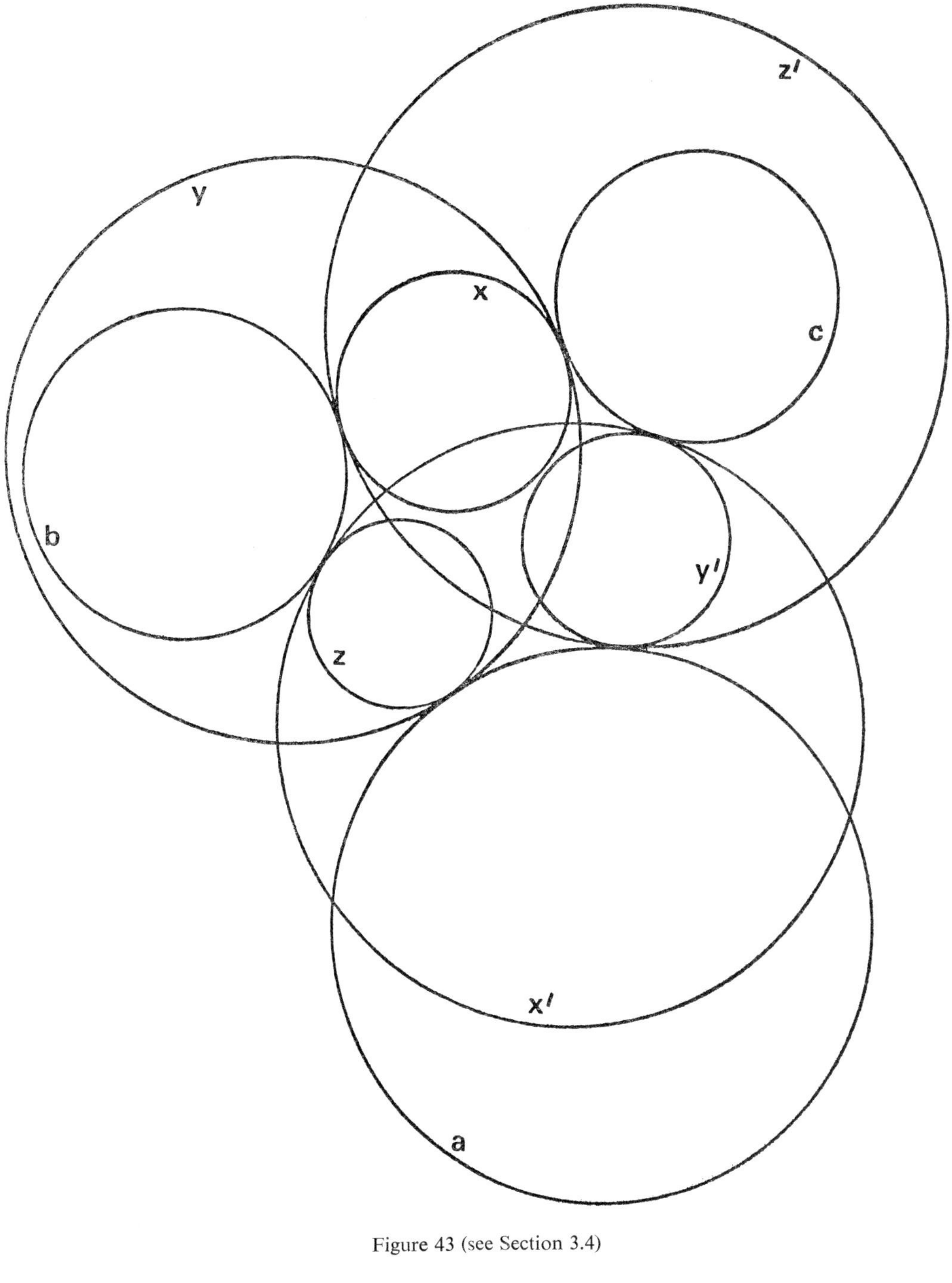

Figure 43 (see Section 3.4)

must be made. We begin by recalling the following facts.

(i) If we are given three circles of general position in the plane, it is possible to draw eight circles to touch all three. When, however, two of the three given circles already touch, as is the case at every stage in constructing the chain of this section, this number is reduced to six (of which two are commonly regarded as 'counting doubly' so as to keep the total number—properly counted—as eight). Thus, for example, there are two circles in the contact coaxal family determined by x and c which also touch a, and these are two possible choices for y in the chain. By analogy with the terminology of the last section, we call these the *special choices* for y (they are the two which count doubly); the other four possible choices for y will be termed *non-special.* A similar distinction between special and non-special choices arises at each subsequent step in the construction of the chain.

(ii) We next note that, if a circle s is drawn to touch two given circles, then the line joining the two points of contact passes through one or other of the two centres of similitude of the given circles and we may, for convenience, say that *s belongs to* that centre of similitude. Thus, of the six possible choices for y in our chain, three choices (one special and two non-special) belong to each centre of similitude of c and a.

(iii) Finally we recall that the centres of similitude of three circles, taken by pairs, lie by threes on four lines (and so are the six vertices of a complete quadrilateral).

With these preliminaries, we can now give a precise statement of the nine circles theorem as follows.

Let a, b and c be three circles in the plane and let us select, once and for all, a set of three collinear centres of similitude (one for each pair of a, b and c). If we then construct chains x, y, . . . of six circles touching pairs of the circles a, b, c in cyclic order and belonging to the chosen centres of similitude (in the technical sense of 'belonging' described in (ii) above) then

(a) if the special choice is made at each stage, the chain will close up; and

(b) if a non-special choice is made at each stage, then x, y, z and x' can be chosen quite arbitrarily and it is then always possible to find 'correct' choices for y' and z' so that the chain closes up.

It will be noted that Figures 41 and 42 illustrate chains of non-special choices whereas Figure 43 shows a special chain.

Proof.

The only comprehensive proof that we know of the nine circles theorem uses the methods of three-dimensional complex projective geometry and does not seem appropriate for insertion in a book such as this.† We shall therefore give here a proof which, though not comprehensive, applies to a large general class of cases of the theorem. Namely, we shall prove the theorem under the assumptions (a) that the three base circles a, b, c have equal radii and (b) that the three chosen centres of similitude are three (collinear) points at infinity.

†This proof has been written up as a paper entitled "A theorem in circle geometry" by J. A Tyrrell and M. T. Powell and was recently published in the *Bulletin of the London Mathematical Society*, Volume 3 (1971) pp. 70-74.

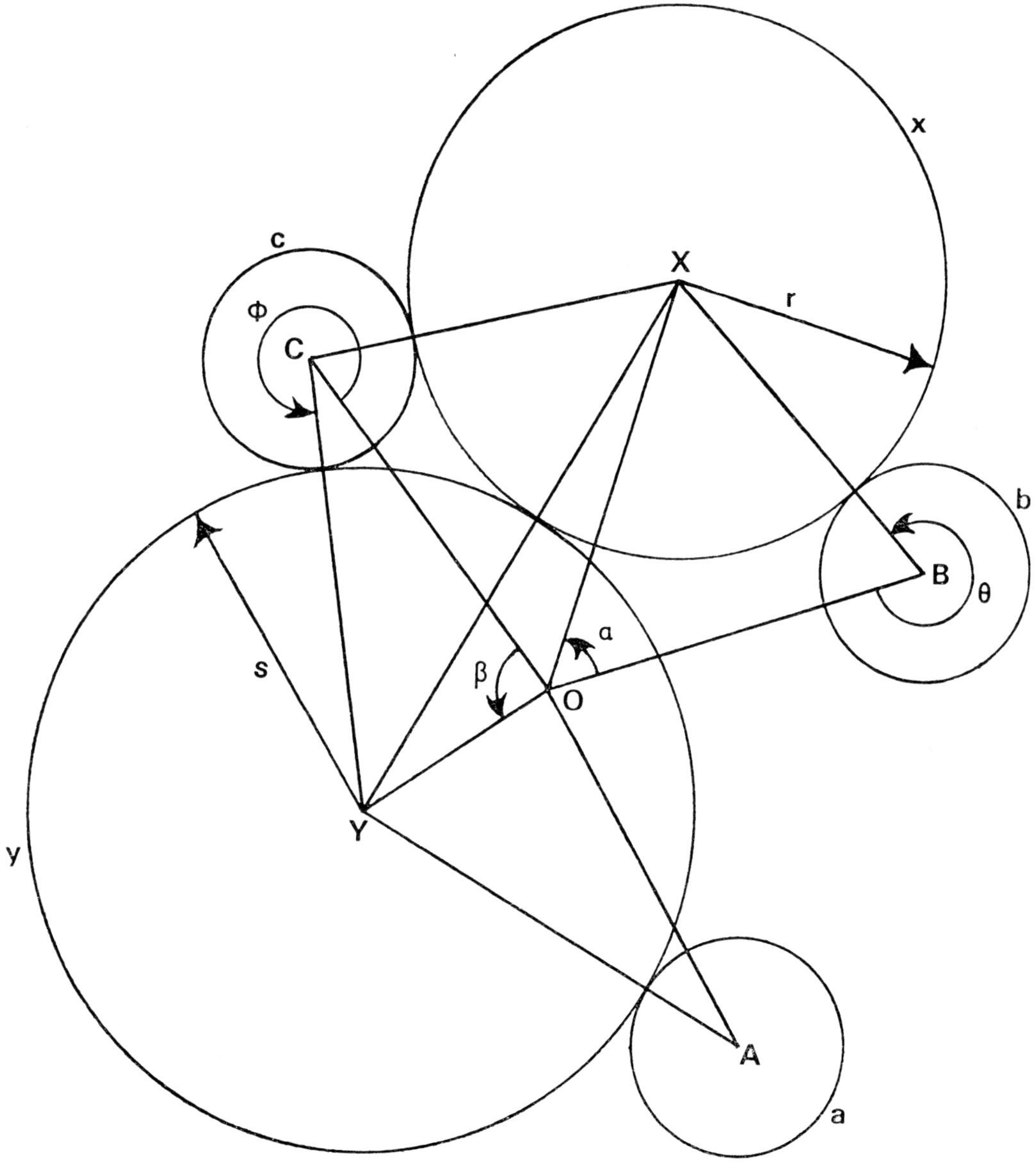

Figure 44 (see Section 3.4)

These, in fact, are not such restrictive assumptions as may at first appear for, as is well-known, three given circles can 'usually' be inverted into three equal circles (where 'usually' means 'provided certain loci intersect in real points') and the chain under discussion is clearly transformed into a similar chain by any inversion.

We consider then (Figure 44) three equal circles a, b, c, with centres A, B, C, and, by suitably choosing the unit of distance, we may assume that these circles have unit radius. We let O be the centre of the circle through A, B and C and denote the radius of this circle by ρ; hence $OA = OB = OC = \rho$. We also let $\angle BOC = 2\alpha$, $\angle COA = 2\beta$ and $\angle AOB = 2\gamma$, noting that $\alpha + \beta + \gamma = \pi$.

Now the circles x which *belong* (in the sense described earlier) to the infinite centre of similitude of b and c are those which touch both b and c externally and those which touch both b and c internally (but not those which touch one internally and the other externally); and the same rule holds for all subsequent circles in the chain. Figure 44 illustrates a position for x (with external contacts) with centre X and radius r together with a non-special choice for y, with centre Y and radius s. We shall denote by θ the angle which BX makes with BO, measured in the positive sense (so that, in our figure, θ is a re-entrant angle) and *we shall regard θ as a parameter which determines the position of x.* We likewise denote by ϕ the angle which CY makes with CO and *regard ϕ as a parameter which determines the position of y.* Similarly, continuing round, we have a parameter ψ (not shown in the figure) which determines the next circle z; θ' which determines x'; ϕ' and ψ' determining y' and z'; and θ'' determining x''; and our problem, of course, is to show that (subject to the right choices being made) $\theta'' = \theta \pmod{2\pi}$.

Let us dispose, first, of the case in which we make a *special* choice at each stage (a case not illustrated by the figure). For this case, y touches c at the same point as x and so (since by symmetry $\angle XBO = \angle XCO$) ϕ is given in terms of θ by $\phi = 2\pi - \theta$. Similarly, continuing round, we get $\psi = 2\pi - \phi$, $\theta' = 2\pi - \psi, \ldots, \theta'' = 2\pi - \psi'$, and it follows at once from these relations that $\theta'' = \theta$, as desired.

The rest of the proof is concerned exclusively with *non-special* choices, and the first step is to obtain the relation between θ and ϕ for this case. We first note that, by symmetry, X lies on the perpendicular bisector of BC and so $\angle BOX = \alpha$. Hence

$$\angle OXB = \pi - \angle BOX - \angle OBX = \theta - \alpha - \pi$$

and, by applying the sine rule to triangle BOX, we obtain

$$\frac{r+1}{\sin\alpha} = \frac{\rho}{\sin(\theta-\alpha-\pi)} \text{ i.e. } \frac{1}{r+1} = \frac{\sin(\alpha-\theta)}{\rho\sin\alpha}. \qquad (1)$$

Similarly, from triangle COY, we obtain

$$\frac{1}{s+1} = \frac{\sin(\beta-\theta)}{\rho\sin\beta}. \qquad (2)$$

Next, noting that $\angle XCY = 4\pi - \theta - \phi$, we have, by the cosine rule in triangle XCY,

$$(r+s)^2 = (r+1)^2 + (s+1)^2 - 2(r+1)(s+1)\cos(4\pi-\theta-\phi)$$

which simplifies easily to

$$\sin^2 \tfrac{1}{2}(\theta+\phi) = \left(1 - \frac{1}{r+1}\right)\left(1 - \frac{1}{s+1}\right).$$

Substituting then from (1) and (2), we obtain

$$\sin^2 \tfrac{1}{2}(\theta+\phi) = \left(1 - \frac{\sin\omega \sin(\alpha-\theta)}{\sin\alpha}\right)\left(1 - \frac{\sin\omega\sin(\beta-\phi)}{\sin\beta}\right) \tag{3}$$

where, for convenience, we have written $\sin\omega$ instead of ρ^{-1}. (N.B. 2ω is the angle between the two tangents from O to any one of the circles a, b, c).

Equation (3), then, is the key relation which gives the parameter ϕ of the circle y in terms of the parameter θ of the circle x. In fact, if θ is given, there are two values of ϕ which satisfy (3), and this corresponds to the fact that, when x is given, there are two non-special choices† for y. Similarly, if we replace θ, ϕ, α and β in (3) by ϕ, ψ, β and γ respectively, we obtain the relation which gives the parameter ψ of the circle z in terms of the parameter ϕ of y; and, if we continue to permute the letters cyclically in this way, we obtain in all six relations analogous to (3) from which we have to eliminate ϕ, ψ, θ', ϕ', ψ' and (hopefully) obtain $\theta'' = \theta \pmod{2\pi}$.

The explicit details of this elimination are rather too lengthy to reproduce here, but we can at least explain the basic trick. Equation (3), as we have already noted, is essentially a two-two relationship between θ and ϕ and, as such, should be solvable by use of elliptic functions. The corresponding equation of Section 3.3—equation (6) of that section—was, by a fortunate chance, a two-two relationship of the special kind that is solvable by use of trigonometric or rational substitutions, and therein lay the reason for the relative simplicity of the proof of the triangle theorem. However, the two-two relationship now under consideration is not of this special kind and genuinely needs an elliptic substitution (or an equivalent device) for its solution. What we look for, then, are suitable new parameters (to replace θ, ϕ, . . .) in terms of which equation (3) reduces to some standard formula of elliptic function theory. We omit the details of how this is contrived, merely stating the outcome as follows.

We introduce, for convenience, the following notations:

$$\lambda = \sqrt{\frac{\sin(\alpha-\omega)}{\sin(\alpha+\omega)}},\quad \mu = \sqrt{\frac{\sin(\beta-\omega)}{\sin(\beta+\omega)}},\quad \nu = \sqrt{\frac{\sin(\gamma-\omega)}{\sin(\gamma+\omega)}}.$$

It does not matter here that λ, μ and ν may not be real, nor which of the two values is taken for each square root; we just make the choice at will in each case. We also write $k = \lambda\mu\nu$.

If, then, we introduce a parameter t given in terms of θ by means of the formula

$$\frac{t-5-k^2}{t+1+5k^2} = \frac{\lambda}{k}\cdot\frac{\sin(\frac{1}{2}\theta+\frac{1}{2}\omega-\frac{1}{4}\pi)}{\sin(\frac{1}{2}\theta-\frac{1}{2}\omega+\frac{1}{4}\pi)} \tag{4}$$

†It should be noticed, incidentally, that if the figure is drawn with x and y in rather different positions, the proof so far may need some modification; but the key relation (3) is obtained always as the end-product.

and if we simultaneously introduce parameters $u, v, \ldots$ related to $\phi, \psi, \ldots$ by means of like formulae with λ replaced by $\mu, \nu, \ldots$ respectively, we find (after a rather long but straightforward calculation) that equation (3)—when expressed as a relation between t and u—comes down to

$$(t+u+f)(4tuf-g_3) = (tu+tf+uf+\tfrac{1}{4}g_2)^2 \tag{5}$$

where $g_2 = 12(1-14k^2+k^4)$, $g_3=8(1+33k^2-33k^4-k^6)$ and

$$f = \frac{(k^2+5)\lambda\mu - (5k^2+1)}{\lambda\mu + 1}.$$

Now equation (5) is a well-known form of the addition theorem for the Weierstrass elliptic function $\wp$ (cf. for example E. T. Copson, *Theory of functions of a complex variable* (Oxford 1935) page 364), and can be written therefore in the form

$$\wp^{-1}(u) \equiv \pm\, \wp^{-1}(t) \pm \wp^{-1}(f) \text{ (mod periods)} \tag{6}$$

where $\wp^{-1}$ denotes the inverse function of $\wp$. (N.B. $\wp^{-1}$ is a two-valued function (mod periods) with equal and opposite values; hence, given a value of t in (6), the right hand side has four values (mod periods), comprising two equal and opposite pairs, and so we obtain two values for u, as expected).

Equation (6), then, expresses the new parameter u of the circle y very simply in terms of the new parameter t of the circle x. There will also be a formula similar to (6) expressing the new parameter v of z in terms of u; indeed, since g_2 and g_3 above were functions of k only, and since k is symmetric in α, β and γ, the relation between v and u is

$$\wp^{-1}(v) \equiv \pm\, \wp^{-1}(u) \pm \wp^{-1}(d) \tag{7}$$

where $\wp$ is the *same* elliptic function as before (and d is a new constant). Likewise, the new parameter t' of x' is given in terms of v by

$$\wp^{-1}(t') \equiv \pm\, \wp^{-1}(v) \pm \wp^{-1}(e) \tag{8}$$

for some other constant e. Combining (6), (7) and (8), we obtain

$$\wp^{-1}(t') \equiv \pm\, \wp^{-1}(t) \pm \wp^{-1}(d) \pm \wp^{-1}(e) \pm \wp^{-1}(f) \tag{9}$$

and this expresses the parameter t' of x' in terms of the parameter t of x. (The sixteen sign alternatives give rise to eight possibilities for t' corresponding to a given t). Similarly, after three more steps, we obtain the parameter t'' of x'' as an eight-valued function of t' in the form

$$\wp^{-1}(t'') \equiv \pm\, \wp^{-1}(t') \pm \wp^{-1}(d) \pm \wp^{-1}(e) \pm \wp^{-1}(f). \tag{10}$$

It is clear now that, if we take the sign alternatives quite arbitrarily in (9), then *it is possible to choose* the sign alternatives in (10) in such a way that, when we eliminate $\wp^{-1}(t')$ between (9) and (10), the terms $\wp^{-1}(d)$, $\wp^{-1}(e)$ and $\wp^{-1}(f)$ cancel off in pairs and leave us with $\wp^{-1}(t'') \equiv \pm\wp^{-1}(t)$, i.e. $t'' = t$ and hence, from (4), $\theta'' = \theta$ (mod 2π). This is the relation that we wished to establish and the proof of our theorem is therefore complete.

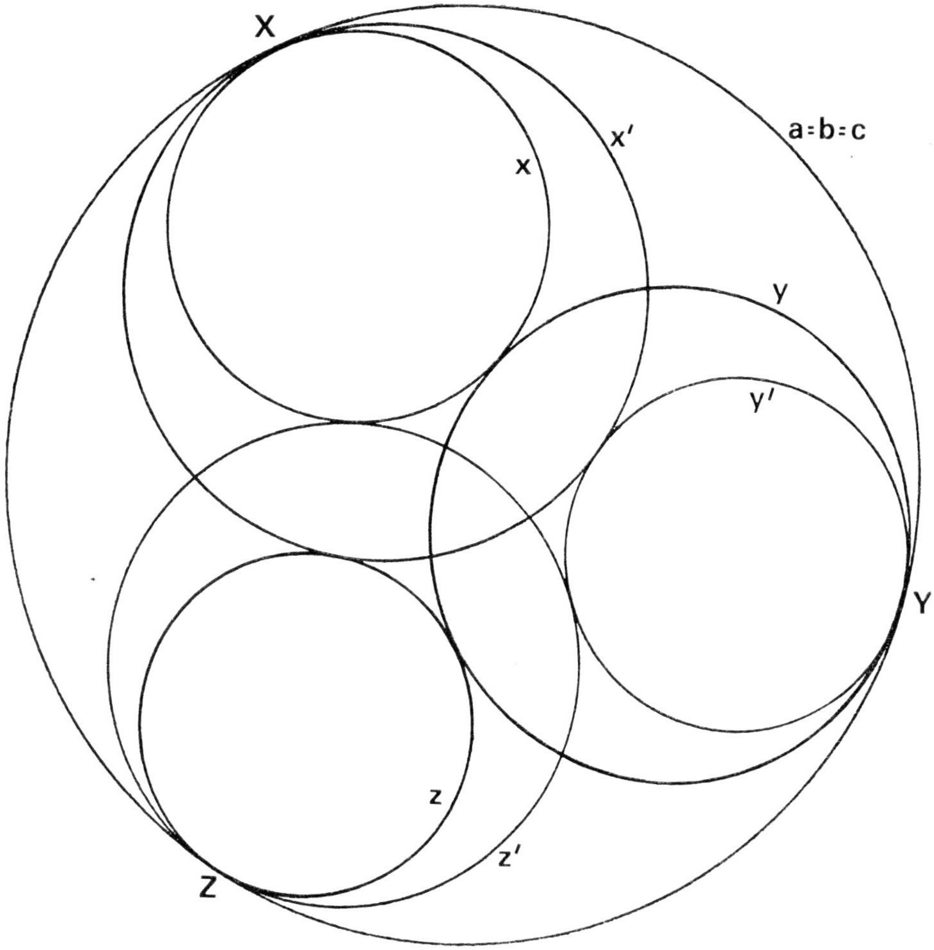

Figure 45 (see Section 3.4)

As the reader will appreciate, the figures for the nine circles theorem come in many possible shapes and sizes, of which only a few have been illustrated in this section. We may now add that the theorem also admits a large number of special cases, obtained by allowing this or that circle to become degenerate. Thus, as we have already seen, the theorem of Section 3.3 is a special case obtained by letting a, b and c degenerate into straight lines. Another obvious special case is obtained by letting a, b and c reduce to point-circles; the proof of this one, incidentally, requires only very elementary arguments and should be within the competence of any schoolboy with a flair for geometry. Yet more special cases can be formed by allowing some of the circles to coincide. Space does not permit a detailed investigation of all these possibilities, but the reader may be interested enough to pursue the matter for himself.

As our parting shot, however, we cannot resist inviting the reader to consider the special form of the nine circles theorem obtained when all three of the circles a, b and c move into coincidence. By drawing figures to illustrate what is happening just before coincidence is reached, it will be discovered that, in the limit, the theorem 'falls apart' into two quite distinct results. One of these is the converse of the seven circles theorem (Section 3.1) while the other asserts the closure of the chain illustrated in Figure 45 (in which the successive circles are required to touch the outer circle at three given points X, Y and Z). We do not claim, however, that this is the best way of proving either of these results, for they can both be proved much more simply by inversion!